Point-of-Care Solution for Osteoporosis Management

Patricia Khashayar

Point-of-Care Solution for Osteoporosis Management

Design, Fabrication, and Validation of New Technology

Springer

Patricia Khashayar
Centre for Microsystems Technology
(CMST)
Faculty of Engineering and Architecture
Ghent University
Ghent, Belgium

ISBN 978-3-319-55052-7 ISBN 978-3-319-55053-4 (eBook)
DOI 10.1007/978-3-319-55053-4

Library of Congress Control Number: 2017933838

Printed on acid-free paper

This Springer imprint is published by Springer Nature
The registered company is Springer International Publishing AG
The registered company address is: Gewerbestrasse 11, 6330 Cham, Switzerland

I dedicate this book to my family; a special feeling of gratitude is due to my loving parents, Mohsen and Shahla, whose words of encouragement and push for tenacity still ring in my ears, and my sister Paria and brother Pouria, who have never left my side and are very special to me.

Preface

Osteoporosis diagnosis, which is nowadays generally made based on bone mineral density (BMD) measurements, suffers from certain limitations. Thus, it is believed that bone turnover markers (BTMs) can help improve osteoporosis detection. While the use of BTMs in osteoporosis diagnosis has certain advantages over BMD and clinical risk assessment tools, not only in monitoring treatment but also in identifying those at risk, the diagnostic value of BTMs in predicting osteoporosis is low, and thus several BTMs are used at the same time with higher accuracy and lower variability to overcome this limitation.

Despite recent advancements in proteomics and personalized medicine, available technologies for BTM measurement, enzyme-linked immunosorbent assay (ELISA) and electrochemiluminescence immunoassay (ECLIA), are time-consuming and laborious and require special labels and expensive instruments. With these apparent limitations, the focus of health-related studies has shifted toward the development of portable, reusable, and at the same time multiplex, label-free point-of-care (PoC) biosensors particularly for diagnosis and monitoring of different diseases. Also known as on-the-spot diagnostic devices, these PoC biosensors should be simple, sensitive, rapid, specific, cheap, and easy to interpret.

Thus, there is a growing demand to fabricate different types of biosensors to provide low-cost miniaturized platforms to assess bone remodeling process more accurately. To our knowledge, none of the existing bone biosensors have succeeded in this regard.

In this book, a microfluidic proteomic platform that can easily be translated into a biomarker diagnostic is reported. This platform integrates microfluidic technology with electrochemical sensing and embodies a reaction/detection chamber to measure serum levels of different biomarkers. The unique design of the platform offers the potential for greater sensitivity as the microfluidic and electrochemical structures can be independently optimized.

This book is the result of a joint Ph.D. in biomedical engineering awarded by Ghent University, Belgium, and the University of Tehran, Iran.

The goal of this work was to develop a reliable, cost-effective, powerful detection and monitoring tool for fracture risk indication and treatment monitoring so that the patient can receive the most appropriate therapy and that physicians can monitor the disease progression.

In this regard, a microfluidic platform capable of measuring serum levels of several biomarkers with high accuracy was developed. The plan was to develop a unique architecture which enabled low sample volume requirements, short assay time, simplicity of fabrication and use, and low-cost instrumentation and at the same time portability and high sensitivity and specificity for the biomarkers. At the end, the measurement results were compared with the current gold standard, electrochemiluminescence immunoassay (ECLIA).

The steps taken to achieve the final goal consisted of:

- The development of a functional conjugated AuNP-antibody nanoprobe to be used to modify the surface of the electrode for sensing purposes: This nanoprobe was directly reacted with antigen following their preincubation in solution, or the probe was assembled onto the electrode to bind with antigen, by so-called "in-solution" or "on-chip" methods, respectively.
- Stable immobilization of the nanoprobe on the electrode, while keeping the antibodies' bioactivity in microreactor environment, and thus supporting high-quality and low-background noise in electrochemical sensing: In this work, gold electrodes were coated with AuNPs by electrodeposition using cyclic voltammetry. This goal is not limited to the development of the modified surface but also includes characterization of the modified surfaces using advanced chemical surface analysis techniques including XPS, STEM, and AFM.
- Validation of the biosensor: One of the most crucial steps in biosensor fabrication is assessing the sensitivity and selectivity of the system. In this work, a calibration curve was plotted for each of the measured markers, and the final results were compared with that of the current state-of-the-art ECLIA.
- Development of microfluidic chips: The application of micro-manufacturing technology and development of microfluidic chips allowed considerable through-put portability and the capacity for a high level of integration and thus fulfilled the requirements of fluidic sample-based point-of-care diagnostics. This also lowered the consumed reagent and sample volume as well as the analysis time without compromising the detection limit of the device, which are required for PoC devices.
- Validation of the final device: In this stage, the parameters which were taken into account to improve the sensitivity and specificity of the microfluidic platform are studied. Hence, the reproducibility of the measurements was investigated through comparing the results with ECLIA.

Ghent, Belgium
January 24, 2017

Patricia Khashayar

Acknowledgments

"To be a doctor is to be an intermediary between man and God," Felix Marti-Ibanez said famously in his To Be a Doctor book. This realization struck a chord with me and made me study medicine. Thanks to my family, I learned that "One of the essential qualities of the clinician is interest in humanity, for the secret of the care of the patient is in caring for the patient." So I dedicated my life to help patients.

As time passed, I struggled with the idea that I could not help all my patients no matter how hard I tried. I imagined that in months and years to come, I would stand by their bedside as patients experienced complications due to late diagnosis or deferred treatment. I couldn't help but strive to seek a solution that would come to their aid.

Adopting the motto "Science is knowing and Technology is doing. Science focuses on understanding the natural phenomena, while technology focuses on understanding the made environment," I realized that bioengineering would provide the perfect blend and present me with the remedy to my internal struggle. My final goal was to incorporate my medical knowledge into developing new devices that would fill the gaps in modern medicine and thus help people around the world.

And thus I started this project, which later become a new beginning and a life changer. I am heartily thankful to Prof. Bagher Larijani, Prof. Jan Vanfleteren and Prof. Ghassem Amoabediny as well as my colleagues Dr. Rik Verplancke and Dr. Frederik Leys for their guidance and persistent help.

And finally, a special thanks to my family. Words cannot express how grateful I am to my mom, my dad, my sister Paria, and my brother Pouria, for all the sacrifices they have made on my behalf. Their prayer for me was what sustained me thus far. I thank them for always taking care of me, for their love and patience, for supporting me all through my life, and for encouraging me to strive toward my goals. They always encourage me to be a better person.

Ghent, Belgium
January 24, 2017

Patricia Khashayar

Acknowledgments

Contents

Acronyms

A

Ab	Antibody
ACA	Anisotropic conductive adhesive
ADC	Analog-to-digital converter
AFM	Atomic-force microscopy
Ag	Antigen
AuNP	Gold nanoparticles
AUX	Counter/auxiliary electrode

B

BALP	Bone alkaline phosphatase
BMD	Bone mineral density
BSA	Bovine serum albumin
BSP	Bone sialoprotein
BTM	Bone turnover marker

C

CA	Chronoamperometry
CatK	Cathepsin K
CE	Capillary electrophoresis
CFD	Computational fluid dynamics
CL	Chemiluminescence
CLIA	Chemiluminescent immunoassay

CNT	Carbon nanotube
COC	Cyclic olefin copolymer
Cr	Creatinine
CRF	Chronic renal failure
CTX	Cross-linked telopeptides of type I collagen
CV	Cyclic voltammetry
C.V.	Coefficient of variation

D

DLS	Dynamic light scattering
Dpd	Deoxypyridinoline
DPV	Differential pulse voltammetry
DTT	Dithiothreitol
DXA	Dual-energy X-ray absorptiometry

E

ECA	Enzymatic colorimetric assay
ECLIA	Electrochemiluminescence immunoassay
ECV	Electrodeposition using cyclic voltammetry
EDC	b-1-Ethyl-3-(3-dimethylaminopropyl) carbodiimide
EIA	Enzyme immunoassay
EIS	Electrochemical impedance spectroscopy
ELISA	Enzyme-linked immunosorbent assay

F

Fab	Fragment antigen-binding region
FET	Field-effect transistor
FITC	Fluorescein isothiocyanate
FTIR	Fourier transform infrared spectroscopy
FRAX	Fracture risk assessment tool
FRF	Frequency response function
Fx	Fracture

G

GnRH	Gonadotropin-releasing hormone
GSH	Glutathione

H

HPLC High-performance liquid chromatography
HRT Hormone replacement therapy
Hyp Hydroxyproline

I

ICMA Immunochemiluminescence assay
IOF International osteoporosis foundation
IRMA Immunoradiometric assay

L

LBL Layer by layer
LLD Lower limits of detection
LoD Limit of detection
LoQ Limit of quantification
LSC Least significant change

M

MEMS Microelectromechanical systems

N

NHS N-hydroxysuccinimide
NP Nanoparticle
NTX N-terminal cross-linking telopeptide of type I collagen

O

Oc Osteocalcin
OPG Osteoprotegerin

P

P1CP Procollagen type 1 C-terminal propeptide
P1NP Procollagen type 1 N-terminal propeptide

PBS Phosphate-buffered saline
PCB Printed circuit board
PDDA Poly(diallyldimethylammonium chloride) solution
PDMS Polydimethylsiloxane
PEG Polyethylene glycol
PMMA Poly(methyl methacrylate)
POC Point of care
PPV Positive predictive value
PTH Parathyroid hormone
PZT Piezoelectric transducer

Q

QCM Quartz crystal microbalance

R

RAI Radioimmunoassay
REF Reference electrode
RSD Relative standard deviation

S

S/N Signal-to-noise ratio
SAM Self-assembled monolayer
SD Standard deviation
SEM Scanning electron microscopy
SPE Screen-printed electrode
STEM Scanning transmission electron microscopy

T

TEM Transmission electron microscopy
Tg Glass transition temperature
Ti Titanium
TRAP Tartrate-resistant acid phosphatase

U

ULD Upper limits of detection
UV Ultraviolet

W

WE Working electrode
WHO World Health Organization

Glossary

Absorbance When a beam of light passes through the sample, the light absorption occurs because of the atoms or molecules taking away the energy from the light. In general, the absorbance of a sample is directly proportional to the optical path length of the sample holder and the concentration of the sample.

Anabolics Anabolic osteoporosis agents such as teriparatide (human parathyroid hormone [1–34]) reduce fracture incidence by improving other bone qualities in addition to increasing bone mass by directly stimulating bone formation.

Antibody Antibody is a large, Y-shaped protein produced mainly by plasma cells that is used by the immune system to identify and neutralize pathogens such as bacteria and viruses. The antibody recognizes a unique molecule of the harmful agent called an antigen. Antibody and antigen interact by spatial complementarity (lock and key).

Antigen An antigen is any substance that causes the immune system to produce antibodies against it. Each antibody is specifically produced by the immune system to match an antigen.

Antiresorptives Antiresorptive osteoporosis agents limit bone loss by decreasing the rate of absorption by osteoclasts, which are cells that absorb bone. By reducing how quickly osteoclasts work, antiresorptives may prevent further loss of bone mass. There are a number of classes of antiresorptives, including bisphosphonates, calcitonin, estrogen therapy or hormone therapy, and estrogen agonists/antagonists.

Biomarker A biomarker is any substance, structure, or process that can be measured in the body or its products and influence or predict the incidence of outcome or disease.

Bone turnover marker Bone turnover markers are bone tissue proteins or their fragments/enzymes released from bone cells during bone turnover. These markers include collagen breakdown products and other molecules released from osteoclasts and osteoblasts during the process of bone resorption and formation.

Calibration Curve A calibration curve is a general method for determining the concentration of a substance in an unknown sample by comparing the unknown to a set of standard samples of known concentration. It shows how the instrumental response, the so-called analytical signal, changes with the concentration of the analyte (the substance to be measured).

Chemiluminescence CL is the light produced by a chemical reaction. The CL intensity is directly proportional to the concentration of a limiting reactant involved in the CL reaction. As a principle, in CL, reactants A and B react with an excited intermediate to form a product and emit the light which can be detected and analyzed by an optical detection system.

Conjugated antibody A conjugated antibody is a monoclonal or polyclonal antibody linked to a label and used for detection in a diverse range of assay techniques.

FET FET is a transistor that uses an electric field to control the shape and electrical conductivity of a charge carrier channel in a semiconductor material. FET's conductivity is regulated by the voltage applied to the gate contact. This voltage imposes an electric field into the device, which in turn attracts or repels charge carriers to or from the region between the source and drain contacts.

Fluorescence Fluorescence is the emission of certain wavelength of visible light from a substance by the absorption of an excitation light of different wavelengths. Fluorescence detection is the most commonly observed form of optical detection, because of its high sensitivity and the ease of integrating a label.

Glutathione GSH is a thiol-containing amino acid with high affinity toward gold nanoparticles through the formation of Au-S bond.

Immunoassay An immunoassay is a biochemical test that measures the presence or concentration of an analyte (usually proteins) in a solution (biological liquids such as serum or urine for medical and research purposes) through the use of an antibody or an antigen.

Limit of detection LoD is the lowest quantity of a substance that can be distinguished from the blank value within a stated confidence limit (generally 1 %). It is estimated from the mean of the blank, the SD of the blank and some confidence factor. Another consideration that affects the LoD is the accuracy of the model used to predict concentration from the raw analytical signal.

Limit of quantification LoQ is the lowest concentration at which the analyte can not only be reliably detected but at which some predefined goals for bias and imprecision are met. The LoQ may be equivalent to the LoD or it could be at a much higher concentration.

Molecular pathway A biological/molecular pathway is a series of actions among molecules in a cell that leads to a certain product or a change in a cell. Such a pathway can trigger the assembly of new molecules, such as a fat or protein. Pathways can also turn genes on and off, or spur a cell to move.

Nanoprobe Nanoprobe refers generically to any chemical or biological technique that deals with nanoquantitles and in other words, that is introducing nanoparticles in aqueous solution to serve in the detection, diagnosis, and treatment of various diseases.

Positive predictive value PPV is the probability that subjects with a positive screening test truly have the disease. In other words, it is the ability of the test to correctly label people who test positive. It is calculated in two ways:

- PPV = (# true positives)/(# true positives + # false positives)
- PPV = sensitivity × prevalence/sensitivity × prevalence + (1 − sensitivity) × (1 − prevalence)

Proteomics Proteomics is a rapidly growing field of molecular biology that is concerned with the systematic, high-throughput approach to protein expression analysis of a cell or an organism. Typical results of proteomics studies are inventories of the protein content of differentially expressed proteins across multiple conditions.

Relative standard deviation RSD is a special form of SD calculated using $\frac{100\times SD}{\bar{x}}$, where SD is the sample standard deviation and $\bar{x}$ is the sample mean.

Surface plasmon resonance When a polarized light hits a prism covered with a thin film of metal at a specific (resonance) angle, SPR occurs. SPR effect is sensitive to the binding of analyte because the increase in mass causes a proportional increase in the refractive index at the interface between a thin metal film and an ambient medium. For an immunoassay, before an antigen binds to an immobilized antibody, the refractive index of the prism is constant. After the binding event, the refractive index changes. In this way, it is possible to monitor the biomolecular interaction on the metal film. Therefore, SPR provides a noninvasive, label-free approach for detecting the concentration of analytes, kinetics, and affinity in real time.

Surrogate marker A surrogate marker is an event or a laboratory value that researchers hope can serve as a reliable substitute for an actual disease. A common example of this is blood cholesterol levels. These levels are surrogates, or substitutes, for heart disease.

T-score T-score is the comparison of a person's bone density with that of a healthy 30-year-old of the same sex at −1.0 and above.

Theranostics Theranostics is used to define ongoing efforts in clinics to develop more specific, individualized therapies for various diseases, and to combine diagnostic and therapeutic capabilities into a single agent.

Upper limits of detection ULD is the largest intrinsic intensity that the source can have and yet have a given probability of remaining undetected.

Voltammetry Voltammetry is an electroanalytical method used in analytical chemistry. In voltammetry, a time-dependent potential is applied to an electrochemical cell, and the resulting current is measured as a function of that potential. The technique provides quantitative and qualitative information about the species involved in the oxidation or reduction reaction.

Chapter 1
Introduction

Abstract What is osteoporosis? Why early diagnosis plays an important role in lowering the burden of this disease on the patient and society? How does a point-of-care device improve osteoporosis management? This chapter deals with the answers to these questions.

1.1 Importance of Point-of-Care Testing in Osteoporosis Management

1.1.1 The Concept of Bone Remodeling

Bone is a dynamically and metabolically active organ that is continuously subjected to resorption and formation by the coordinated action of the two different categories of bone cells, namely osteoclasts (cells responsible for resorbing bone), and osteoblasts (cells responsible for building bone) and their descendants osteocytes. These two processes are collectively called bone turnover or bone remodeling and they are coupled in time and space [1] (Fig. 1.1). After middle age, bone loss occurs as resorption exceeds formation. This process is known as osteoporosis [2].

1.1.2 What Is Osteoporosis?

Osteoporosis, literally meaning "porous bone," is the most frequently found metabolic bone disease (Fig. 1.2). The systemic skeletal disease is characterized by

- Low bone mass
- Microarchitectural deterioration of the bone
- Increased bone fragility and fracture risk.

Osteoporosis has been operationally defined on the basis of bone mineral density (BMD) assessment. Using the operational definition of the World Health Organization (WHO) and based on the BMD measurements at the spine, hip, or forearm by dual X-ray absorptiometry (DXA) devices, individuals are divided into three groups according to their bone mass status [3]:

P. Khashayar, *Point-of-Care Solution for Osteoporosis Management*,
DOI 10.1007/978-3-319-55053-4_1

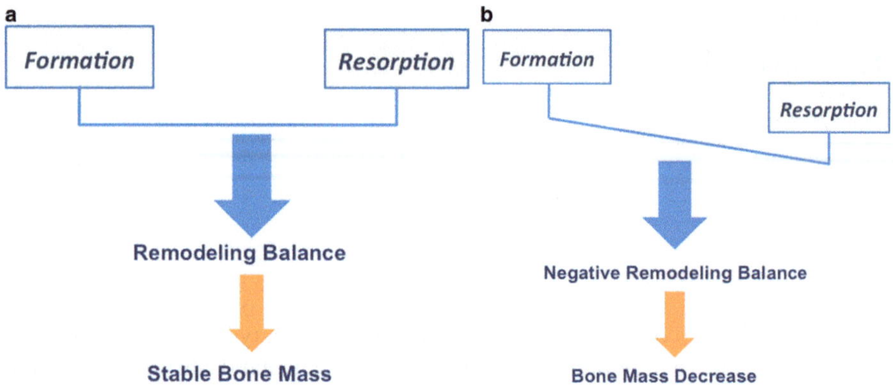

Fig. 1.1 (**a**) Formation and resorption status leading to stable bone mass. (**b**) Formation and resorption status leading to bone loss

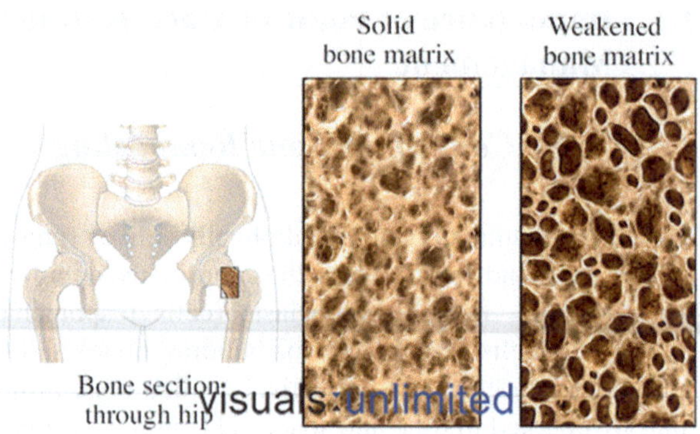

Fig. 1.2 Healthy (L) and osteoporotic (R) bone (same scale)

- Normal: BMD lies within 1 standard deviation (SD) of the average value for young healthy adults (T-score at −1.0 and above).
- Low bone mass (osteopenia): BMD lies between 1.0 and 2.5 SD below that of the average value for young healthy adults (T-score between −1.0 and −2.5).
- Osteoporosis: BMD lies 2.5 SD or more below that of the average value for young healthy adults (T-score at or below −2.5).

1.1.2.1 Epidemiology of Osteoporosis

Osteoporosis is a multifactorial systemic disease that affects a large number of people in different parts of the world (Fig. 1.3). It is more prevalent than breast and prostate cancer, which are considered as the most common tumors in each gender.

Osteoporosis is currently considered as a serious global public health concern, with over 200 million people suffering from the disease worldwide. According to the WHO criteria, osteoporosis affects 30% of postmenopausal white women in the USA (9.4 million women) [4].

Fig. 1.3 Worldwide risk of developing osteoporosis in different decades

The prevalence of osteoporosis in the EU was estimated at 27.6 million in 2010 [5]. In Belgium, 120,695 men aged over 50 (6.5%) and 476,875 women of the same age group (22.4%) were reported to suffer from osteoporosis in 2010.

Latest figures show that 25,969,046 and 3,024,798 Iranians suffer from osteopenia and osteoporosis, adding that the rate would experience a rapid surge and reach 40,303,730 and 3,592,708 by 2020 [6] (Fig. 1.4).

1.1.2.2 Osteoporosis Risk Factors

Despite the common misperception that osteoporosis is always the result of bone loss due to aging, reports have shown that failing to reach optimal (i.e., peak) bone mass during childhood and adolescence may also result in osteoporosis in the absence of accelerated bone loss [7]. Hence sub-optimal bone growth in childhood and adolescence plays as important role as bone loss in the development of osteoporosis.

A number of factors can increase the likelihood of developing osteoporosis; these factors are divided into two main groups (Table 1.1).

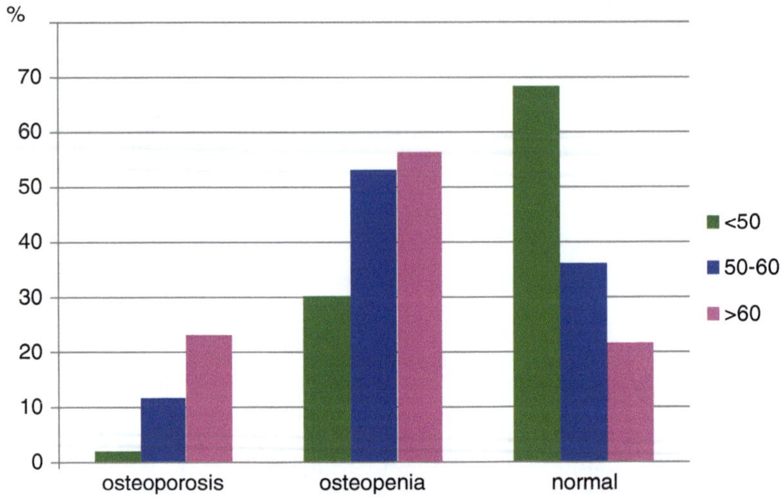

Fig. 1.4 Number of Iranians suffering from different degrees of bone loss

Table 1.1 Risk factors attributed to osteoporosis

Major risk factors	Minor risk factors
Age ≥ 65 years	Rheumatoid arthritis
Family history of osteoporotic fracture	Past history of clinical hyperthyroidism
Fragility fracture after age 40	Chronic anticonvulsant therapy
Vertebral compression fracture	Low dietary calcium intake
Apparent osteopenia on X-ray film	Smoking
Malabsorption syndrome	Excessive alcohol intake
Primary hyperparathyroidism	Excessive caffeine intake
Propensity to fall	Weight < 57 kg
Systemic glucocorticoid therapy for > 3 months	Weight loss > 10% of weight at age 25
Hypogonadism	Chronic heparin therapy
Early menopause (before age 45)	

1.1.2.3 Complications of Osteoporosis

Osteoporosis is often called a "silent disease" as bone loss occurs without any symptoms. The affected individual, therefore, would not become aware of having the condition until their bones become so weak that a sudden strain or fall causes fracture or vertebral collapse. Its major complication, fracture, is associated with diminished life-quality, shortened lifespan, and heavy health care costs [8]. In addition to morbidity, osteoporotic fractures also increase mortality risk in the affected patients [9]. Reports estimate that after the age of 50, nearly 40% of women and 13% of men in western countries will sustain a fragility fracture [4] (Fig. 1.5).

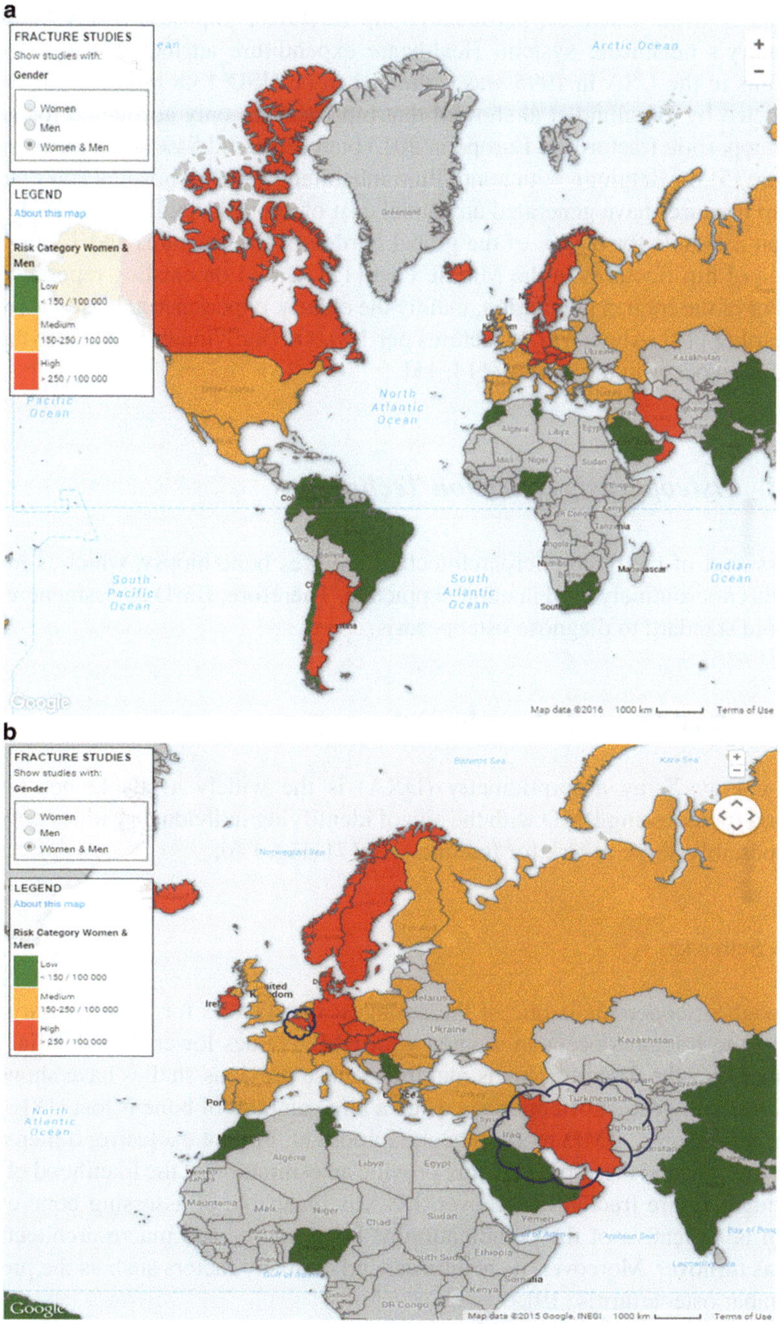

Fig. 1.5 (**a**) Hip fracture incidence map from International Osteoporosis Foundation (https://www.iofbonehealth.org/facts-and-statistics/hip-fracture-incidence-map). Reprinted with permission. (**b**) Iran and Belgium have a high risk for hip fracture in both genders

This is while fractures, particularly hip fractures, impose a heavy burden on a country's healthcare system. Healthcare expenditure attributed to osteoporotic fractures in the USA in 1995 was estimated to be USD 13.8 billion [10]. A study conducted by Hernlund et al showed that hip fractures only accounted for 17% of all osteoporotic fractures in Europe in 2013 but comprised 54% of the direct cost of fracture [5]. In Belgium, with ten million inhabitants, 13,150 hospital stays per year for hip fractures have generated an annual cost of about 118 million USD [11].

Iran accounts for 0.85% of the global burden of hip fractures and 12.4% of the burden of hip fractures in the Middle East [12]. Based on existing reports, a large number of the Iranian population, mainly the elderly, experience at least one fracture in their life [13]. About 472.1 fractures per 100,000 (individuals aged more than 50) population occur in the country [14, 15].

1.1.3 Osteoporosis Detection Techniques

Assessment of the bone microarchitecture requires bone biopsy, which is invasive and thus not routinely used in clinical practice. Therefore, BMD assessment remains the gold standard to diagnose osteoporosis.

1.1.3.1 Dual-Energy X-Ray Absorptiometry

Dual-energy X-ray absorptiometry (DXA) is the widely available non-invasive method for assessing BMD with the aim of identifying individuals with osteoporosis and, possibly, those at-risk for fracture [16, 17] (Fig. 1.6).

DXA Setbacks

Although osteoporosis is one of the strongest risk factors for fracture, only about half of the fractures occur in women with BMD values lower than the diagnostic threshold for the disease. This is mainly because previous studies have shown that BMD can only detect osteoporosis when a large amount of bone is lost [18].

In other words, BMD represents one important, but not exclusive, dimension of bone strength and therefore can only provide an estimation of the likelihood of experiencing a future fracture. Moreover, DXA is incapable of assessing bone quality, which is indicative of the combination of bone micro- and macro-architecture as well as turnover. Moreover, its results depend on many factors such as the presence of lumbar osteoarthritis [19].

On the other hand, while BMD changes are considered as a valid end point for verifying the efficacy of osteoporosis treatments, BMD changes and fracture risk reduction rate following medication use is not measurable at early stages [20]. It takes more than two years for DXA to be able to detect changes in bone mass and thus possible response to treatment [21].

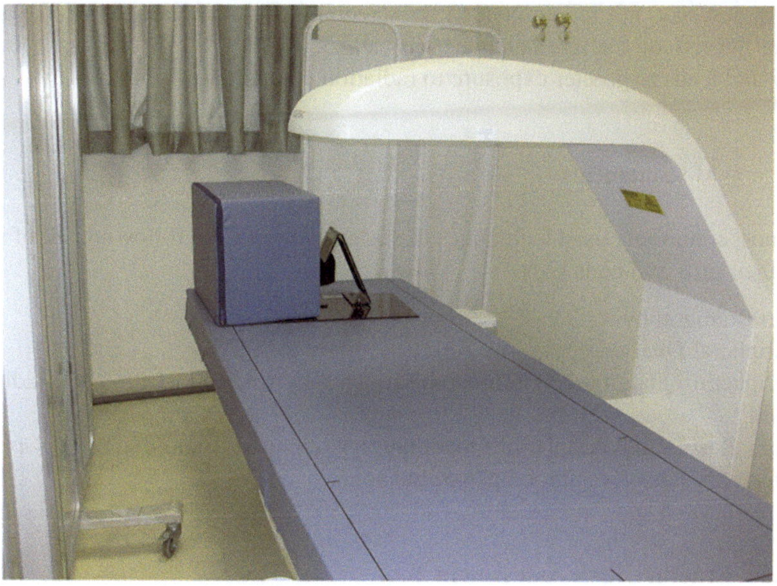

Fig. 1.6 Dual-energy X-ray absorptiometry (DXA) device

Furthermore, BMD alone is incapable of diagnosing patients suffering from secondary osteoporosis, caused by certain underlying medical conditions or the use of medications that may cause bone loss [7]. In other words, bone loss could be secondary to an increase in bone resorption (due to postmenopausal state, secondary hyperparathyroidism, high inflammatory status, etc.) or decrease in bone formation (due to steroid use, adynamic bone disorders due to hypoparathyroidism and/or vit-D deficiency, etc.). Osteoporosis management in these patients is therefore based on the treatment of the underlying cause and mostly requires additional measurement of certain markers including bone turnover markers (BTMs).

In addition, there have been major advances in the number and range of agents available for osteoporosis treatment, all with proven anti-fracture efficacy [22]. These agents have differing modes of action and these need to be taken into account when developing monitoring strategies. Ideally, only those individuals who are at high risk of sustaining a fracture should receive anabolic or antiresorptive medication and treatment, should be targeted on the basis of risk assessment rather than the BMD measurement alone.

1.1.3.2 Other Imaging Modalities

Quantitative Computerized Tomography

QCT measures volumetric bone density of the spine and hip and analyzes cortical and trabecular bone separately [23].

It is, however, not recommended for screening as the application of T-scores to predict the risk of fracture has not been validated with QCT. Moreover, it is more costly and leads to greater exposure to radiation compared with DXA.

Peripheral Measurements

The most commonly used techniques in this category are the following, all of which have their pros and cons [24]:

- Ultrasonography
- Peripheral Dual-energy X-ray Absorptiometry (pDXA)
- Peripheral Quantitative Computed Tomography (pQCT) of the heel, radius, or hand
- High Resolution Peripheral Quantitative Computed Tomography (HR-pQCT), which provides a virtual biopsy of the bone.

1.1.3.3 Fracture Risk Assessment Tool

In an attempt to better estimate fracture risk even when BMD measurement is not available, WHO developed the FRAX tool, which can calculate the 10-year probability of osteoporotic fracture based on clinical risk factors [25]. The introduction of the multivariate algorithm-based fracture risk assessment tool has evolved the definition of the disease, leading to treatment decisions based on absolute risk of fracture threshold. The application of the tool in clinical practice has resulted in a higher positive predictive value (PPV), reduced number of individuals needed to be treated to prevent one fracture and more importantly enhanced sensitivity [26].

There are, however, a number of general and methodology-specific limitations for FRAX, as the tool only evaluates nine of the many risk factors related to osteoporosis and more importantly does not consider all the medications which influence fracture risk. Moreover, in this model, risk factors are quantified in a binary approach.

QFracture and Garvan are among other existing fracture risk assessment tools [27]. QFracture includes a larger number and wider variety of clinical risk factors compared with FRAX and GARVAN, and is reported to be more accurate. But still FRAX is considered as the most validated instrument in fracture prediction.

1.1.4 Early Diagnosis in Osteoporosis Detection

Osteoporosis, once thought to be a natural part of ageing among women, is no longer considered age or gender-dependent. It is largely preventable due to the remarkable progress in the scientific understanding of its causes, diagnosis, and treatment [28].

Optimization of bone health is a process that must occur throughout the lifespan in both genders. Factors that influence bone health in different age groups are essential to prevent osteoporosis and its consequences [29].

Many studies therefore have expressed the importance of prevention and therefore early detection before any fracture has happened in reducing the burden of the disease [9]. According to these studies, in view of the fact that the restoration of diminished bone volume seems quite difficult, early diagnosis of osteoporosis is the most valuable strategy for preventing osteoporotic fractures.

Despite recent advances in diagnosis of osteoporosis, assessment of fracture risk, development of interventions that reduce risk of fractures, and the development of practical guidelines, surveys indicate that only a minority of men and women at high fracture risk actually receive treatment [30]. This is while many osteoporosis guidelines recommend intervention to be considered for men and women who have sustained a fragility fracture, stressing that these individuals are at an increased risk of a second fracture [31].

Recent statistics show that less than 20% of patients who sustain a fragility fracture receive treatment to reduce the risk of future fracture within the year following the first one [32]. The therapeutic care gap is even wider in the elderly in whom the importance and impact of treatment is well known. Studies have shown that as few as 10% of older women with fragility fractures receive any osteoporosis therapy (estrogens not considered) [31].

Early diagnosis can help to not only identify those who would benefit from intervention but also determine the optimal manner in which response to treatment should be monitored.

1.2 The Aim of Current Book

Current trends towards theranostics and provision of personalized diagnostic therapy tailored to an individual have emphasized the need for inexpensive point-of-care (PoC) devices, capable of performing rapid analysis, with small volumes of sample, minimum number of assay steps, and no need for highly skilled personnel for routine check up and patient screening. PoC diagnostics are analytical testing platforms that can be used both in a clinical laboratory and in the vicinity of a patient using portable equipment [33].

Therefore, the entire clinical diagnosis and treatment community is witnessing a paradigm shift from conventional diagnostic devices to the miniaturized and PoC versions [34]. These devices can fulfill the increasing demand of the medical sector for fully automated instruments that can directly use unprocessed specimens and that require minimal electronic or mechanical maintenance [35].

PoC diagnostics have the potential to provide rapid and accurate results, at reasonable costs. An ideal diagnostic device should provide a suitable environment for interaction of the analyte to be tested in a given sample with the bio-recognition molecule (e.g., protein, lipid, nucleic acid, whole cell), which is immobilized onto the sensor surface.

PoC diagnostics are critically dependent on the combination of the analytical tools and the rapidly emerging knowledge on molecular biomarkers. They can be used to identify both physiological and biological entities associated with diseases, and have evolved from simple single physiological parameters to novel markers that span fields from genomics to metabolomics.

1.2.1 PoC and Osteoporosis

As mentioned earlier, the restoration of diminished bone volume is difficult to achieve and thus early diagnosis of osteoporosis is the most valuable strategy in preventing osteoporotic fractures. Prevention of fracture starts by early identification of fracture-prone individuals [36]. However, the prediction of rapid bone loss, incidence of osteoporosis, and osteoporotic fracture remains difficult. The identification of individuals who would best benefit from intervention and for those on treatment, the optimal manner in which response to treatment should be monitored are therefore important gaps in the clinical armamentarium in osteoporosis management.

Recent studies have pointed out the importance of BTMs in this regard [37]. BTMs are bone tissue proteins or their fragments/enzymes released from bone cells during bone turnover.

Bone remodeling rate therefore can be assessed by measuring BTM levels in serum or urine. A more detailed explanation of BTMs and their characteristics could be found in Chap. 3. Biochemical markers of bone turnover have been investigated in osteoporosis for three main purposes:

- to monitor anabolic or antiresorptive therapy,
- to predict potential bone loss and the risk of developing osteoporosis,
- to improve assessment of fracture risk in individuals.

Although BTMs have shown clinically interesting association with bone loss rate, the routine use of BTMs in clinical practice is not generally recommended, as these tests vary in sensitivity and specificity [38].

This is while enzyme-linked immunosorbent assay (ELISA), electrochemiluminescence immunoassay (ECLIA), and other prevailing laboratory techniques in the current state of the art for biomarker quantification do not lend themselves well to miniaturized automated systems for application at PoC [39]. Moreover, they are time consuming, expensive, labor intensive, require trained personnel, and lack consistency in laboratory accuracy. Thus performing rapid multiplexed measurements to monitor several biochemical parameters at the same time to reach a rapid and accurate medical decision still remains a largely unmet challenge.

Biosensors, on the other hand, can help determine reaction kinetics of marker interaction in real-time. The technique needs only one step (reaction) and can thus be used for point-of-care screening. As a result, several biosensors were developed during the past years to improve measurement of some of these markers, none of them are routinely used in practice. Detailed explanation of these sensors is discussed in Chap. 3.

In this context, multiplexed detection of different biomarkers using a biosensor has potential to play a very influential role in diagnostics; and advances in microfluidic and nanomaterial technologies have paved the way for the development of lab-on-a-chip PoC instruments [40]. Microfluidics is an extremely attractive technology because of its ability to manipulate small volumes of fluid (micro- to nanoliters) within given spatial dimensions, and the ease of multiplexing and integrating this system into low cost, compact, and portable devices for on-chip testing [41]. It, therefore, provides a very promising way to miniaturize immunoassays, which has played an important role in the diagnosis of different diseases in the past 50 years. Such devices also provide added benefits such as portability, increased reliability, improved sensitivity, decreased analysis time, minimal reagent consumption, and parallel processing [42, 43]. Among different techniques used to generate a signal in these platforms, the electrochemical method has shown the most suitable results because of its simple instrumentation and easy signal quantification [44]. A clear challenge resides in the integration and operation of such microfluidic platforms for measurement of different biomarkers, such as BTMs.

In this book, we describe the design and fabrication of a microfluidic proteomic platform that can easily be translated into a biomarker diagnostic. This platform integrates microfluidic technology with electrochemical sensing and embodies a reaction/detection chamber to measure serum levels of different biomarkers. The unique design of the platform offers microfluidic and electrochemical structures to be independently optimized.

We demonstrate the implementation, feasibility, and specificity of a platform (Osteokit) for accurate measurement of serum levels of BTMs using osteocalcin (Oc) and cross-linked Telopeptides of type I collagen (CTX). The goal was to develop a reliable, cost-effective, powerful detection and monitoring tool for fracture risk indication, and treatment monitoring so that the patient can receive the most appropriate therapy and that physicians can monitor the disease progression.

In other words, this book explains the first two steps, "Assay Development" and "Analytical Validation," before a newly developed "Biomarker Assay" enters the market (Fig. 1.7).

This platform serves as the basis for a clinical diagnostic that can be used by healthcare providers to identify patients at-risk of osteoporotic fracture, those who may benefit more from treatment and monitor treatment effectiveness. The use of microfluidic components together with electrochemical detection suggests that the proposed device could be operated as a hand-held device in future.

1.3 Organization of the Book

The final goal of this book is to develop a microfluidic platform capable of measuring serum levels of several biomarkers with high accuracy. The plan was to develop a unique architecture, which enabled low sample volume requirements, short assay time, simplicity of fabrication and use, and low cost instrumentation

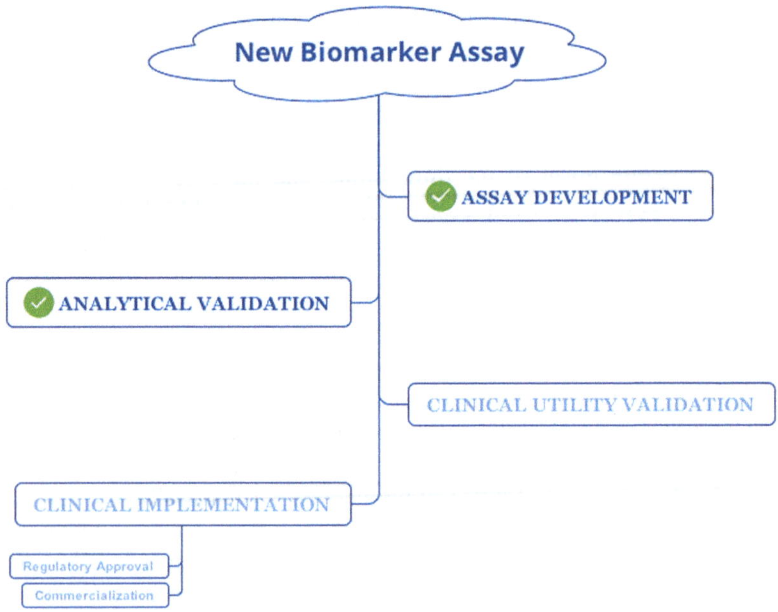

Fig. 1.7 New biomarker assay development steps

and at the same time portability, high sensitivity and specificity for the biomarkers (Fig. 1.8). Furthermore, we wanted our measurement results to be comparable in sensitivity with the current gold standard ECLIA.

Here, we propose the development of a functional conjugated AuNP-antibody nanoprobe to be used to modify the surface of an electrode for sensing purposes as the first step. This nanoprobe will directly react with antigen following their pre-incubation in solution or the probe will be assembled onto the electrode to bind with antigen, by the so-called in-solution or on-chip methods, respectively.

The book is divided into nine chapters.

- Chapter 1 consists of an introduction, explaining the importance of point-of-care (PoC) testing in osteoporosis management and the application field of this work followed by the scientific goals and strategy.
- Chapter 2 consists of a literature overview on BTMs and existing bone biosensors. In the next section, special attention is paid to methods used to deposit gold nanoparticles on a gold electrode, and protein immobilization techniques.
- Chapter 3 covers scientific information and background of the major scientific subjects of this thesis. The first part focuses on biosensors and then basic principles of electrochemistry and microfluidic devices are provided.
- Chapter 4 includes the verification of the conjugated AuNP-antibody nanoprobe (in-solution nanoprobe) fabricated for electrochemical biosensors.
- Chapter 5 explains the fabrication process of the electrodes used in this book. Stable immobilization of the nanoprobe on the electrode, while keeping the

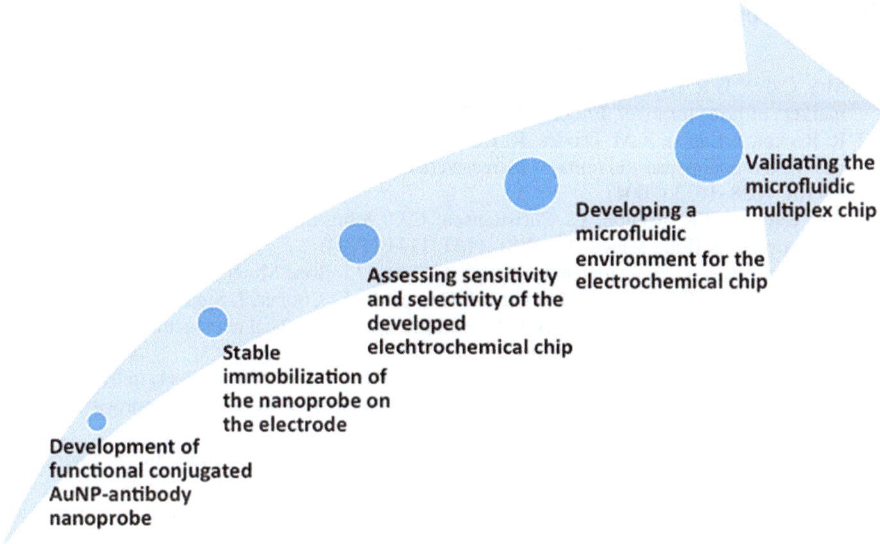

Fig. 1.8 Workflow

antibodies' bioactivity in microreactor environment, and thus supporting high quality and low-background noise in electrochemical sensing was the next goal. Different modification methods to deposit AuNPs on gold electrodes are therefore compared in this chapter. This explanation is not limited to the development of the modified surface using various techniques but also includes characterization of the modified surfaces using advanced chemical surface analysis techniques including XPS, STEM, and AFM.

- Chapter 6 discusses characteristics and validation process of the Oc and CTX biosensors. One of the most crucial steps in biosensor fabrication is assessing the sensitivity and selectivity of the system. In this chapter a calibration curve was plotted for each of the measured markers and the final results were compared with that of the current state-of-the-art ECLIA.
- Chapter 7 discusses the Osteokit prototype fabrication technology. The application of micro-manufacturing technology and development of microfluidic chips allow considerable throughput portability and the capacity for a high level of integration, and thus helps fulfill the requirements of fluidic sample-based point-of-care diagnostics. This also lowers the consumed reagent and sample volume as well as the analysis time without compromising the detection limit of the device, which are required for PoC devices.
- Chapter 8 reports the calibration curve plotted for the final device. Moreover, the accuracy of the measurements was investigated through comparing the results with ECLIA. In this stage, the parameters which were taken into account to improve the sensitivity and specificity of the microfluidic platform are also studied.
- Chapter 9 consists of general conclusions and future prospects of the performed research.

References

1. M.S. Calvo, D.R. Eyre, C.M. Gundberg, Molecular basis and clinical application of biological markers of bone turnover. Endocr. Rev. **17**(4), 333–368 (1996)
2. R. Recker, J. Lappe, K.M. Davies, R. Heaney, Bone remodeling increases substantially in the years after menopause and remains increased in older osteoporosis patients J. Bone Miner. Res. **19**(10), 1628–1633 (2004)
3. J.A. Kanis, L.J.R. Melton, C. Christiansen, C.C. Johnston, N. Khaltaev, The diagnosis of osteoporosis. J. Bone Miner. Res. **9**(8), 1137–1141 (1994)
4. L. Melton, How many women have osteoporosis now? J. Bone Miner. Res. **10**, 175–177 (1995)
5. E. Hernlund, A. Svedbom, M. Ivergård, J. Compston, C. Cooper, J. Stenmark, E. McCloskey, B. Jönsson, J. Kanis, Osteoporosis in the European union: medical management, epidemiology and economic burden. Arch. Osteoporos. **8**, 136 (2013)
6. B. Larijani, H. Resch, J. Bonjour, H.A. Meybodi, M.M. Tehrani, Osteoporosis in Iran, overview and management. Iran. J. Public Health. A Supplementary Issue on Osteoporosis, pp. 1–13 (2007)
7. K.D. Harper, T.J. Weber, Secondary osteoporosis. diagnostic considerations. Endocrinol. Metab. Clin. N. Am. **27**(2), 325–348 (1998)
8. C. Cooper, G. Campion, L.J. Melton, Hip fractures in the elderly: a world-wide projection. Osteoporos. Int. **2**, 285–289 (1992)
9. Scientific Advisory Board, Osteoporosis Society of Canada, Clinical practice guidelines for the diagnosis and management of osteoporosis. CMAJ **155**(8), 1113–1133 (1996)
10. N. Ray, J.K. Chan, M. Thamer, L. Melton, Medical expenditures for the treatment of osteoporotic fractures in the united states in 1995: report from the national osteoporosis foundation. J. Bone Miner. Res. **12**, 24–35 (1997)
11. J. Reginster, P. Gillet, S.W. Ben, G. Brands, O. Ethgen, C. de Froidmont, C. Gosset, Direct costs of hip fractures in patients over 60 years of age in Belgium. Pharmacoeconomics **15**(5), 507–514 (1999)
12. S. Ahmadi-Abhari, A. Moayyeri, F. Abolhassani, Burden of hip fracture in Iran. Calcif. Tissue Int. **80**(3), 147–153 (2007)
13. F. Hadaegh, H. Harati, A. Ghanbarian, F. Azizi, Prevalence of coronary heart disease among Tehran adults: Tehran lipid and glucose study. East. Mediterr. Health J. **15**(1), 157–166 (2009)
14. S. Ghafoori, A. Keshtkar, P. Khashayar, M. Ebrahimi, M. Ramezani, Z. Mohammadi, F. Saeidifard, N. Nemati, M. Khoshbin, S. Azizian, F. Zare, S. Shirazi, B. Larijani, The risk of osteoporotic fractures and its associating risk factors according to the FRAX model in the Iranian patients: a follow-up cohort. J. Diabetes Metab. Disord. **13**, 93 (2014)
15. I. Salehi, S. Khazaeli, S. Najafzadeh, H. Ashraf, M. Malekpour, High prevalence of low bone density in young Iranian healthy individuals. Clin. Rheumatol. **28**, 173–177 (2009)
16. J. Kanis, O. Johnell, C. De Laet, B. Jonsson, A. Oden, A. Ogelsby, International variations in hip fracture probabilities: implications for risk assessment. J. Bone Miner. Res. **17**(7), 1237–1244 (2002)
17. K. Briot, C. Roux, What is the role of dxa, qus and bone markers in fracture prediction, treatment allocation and monitoring? Best Pract. Res. Clin. Rheumatol. **19**(6), 951–964 (2005)
18. S. Schuit, M. van der Klift, A. Weel, C. de Laet, H. Burger, E. Seeman, A. Hofman, A. Uitterlinden, J. van Leeuwen, H. Pols, Fracture incidence and association with bone mineral density in elderly men and women: the Rotterdam study. Bone **34**(1), 195–202 (2004)
19. C.A. Cefalu, Is bone mineral density predictive of fracture risk reduction? Curr. Med. Res. Opin. **20**(3), 341–349 (2004)
20. S.L. Bonnick, L. Shulman, Monitoring osteoporosis therapy: bone mineral density, bone turnover markers, or both? Am. J. Med. **119**(4), Suppl 1, S25–S31 (2006)
21. E. Siris, Y. Chen, T. Abbott, E. Barrett-Connor, P. Miller, L. Wehren, M. Berger, Bone mineral density thresholds for pharmacological intervention to prevent fractures. Arch. Intern. Med. **164**(10), 1108–1112 (2004)

22. V. Elliot-Gibson, E. Bogoch, S. Jamal, D. Beaton, Practice patterns in the diagnosis and treatment of osteoporosis after a fragility fracture: a systematic review. Osteoporos. Int. **15**(10), 767–778 (2004)
23. G. Guglielmi, Quantitative computed tomography (QCT) and dual x-ray absorptiometry (DXA) in the diagnosis of osteoporosis. Eur. J. Radiol. **20**(3), 185–187 (1995)
24. G. Guglielmi, S. Muscarella, A. Bazzocchi, Integrated imaging approach to osteoporosis: state-of-the-art review and update. RadioGraphics **31**(5), 1343–1364 (2011)
25. J.A. Kanis, O. Johnell, A. Oden, H. Johansson, E. McCloskey, FRAX and the assessment of fracture probability in men and women from the UK. Osteoporos. Int. **19**(4), 385–397 (2008)
26. H. Johansson, J.A. Kanis, A. Oden, O. Johnell, E. McCloskey, BMD, clinical risk factors and their combination for hip fracture prevention. Osteoporos. Int. **20**(10), 1675–1682 (2009)
27. A. Marques, R.J.O. Ferreira, E. Santos, E. Loza, L. Carmona, J.A.P. da Silva, The accuracy of osteoporotic fracture risk prediction tools: a systematic review and meta-analysis. Ann. Rheum. Dis. **74**(11), 1958–1967 (2015)
28. J.T. Lin, J.M. Lane, Osteoporosis: a review. Clin. Orthop. Relat. Res. **425**, 126–134 (2004)
29. N. Zethraeus, F. Borgström, O. Ström, J. Kanis, B. Jönsson, Cost-effectiveness of the treatment and prevention of osteoporosis—a review of the literature and a reference model. Osteoporos. Int. **18**(1), 9–23 (2007)
30. L. Jennings, A. Auerbach, J. Maselli, P. Pekow, P. Lindenauer, S. Lee, Missed opportunities for osteoporosis treatment in patients hospitalized for hip fracture. J. Am. Geriatr. Soc. **58**, 650–657 (2010)
31. J.A. Kanis, A. Svedbom, N. Harvey, E. McCloskey, The osteoporosis treatment gap. J. Bone Miner. Res. **29**(9), 1926–1928 (2014)
32. A. Diez-Perez, F.H. Hooven, J.D. Adachi, S. Adami, F.A. Anderson, S. Boonen, R. Chapurlat, J.E. Compston, C. Cooper, P. Delmas, S.L. Greenspan, A.Z. Lacroix, R. Lindsay, J.C. Netelenbos, J. Pfeilschifter, C. Roux, K.G. Saag, P. Sambrook, S. Silverman, E.S. Siris, N.B. Watts, G. Nika, S.H. Gehlbach, Regional differences in treatment for osteoporosis. the global longitudinal study of osteoporosis in women (glow). Bone **49**(3), 493–498 (2011)
33. J. Wang, Electrochemical biosensors: towards point-of-care cancer diagnostics. Biosens. Bioelectron. **21**(10), 1887–1892 (2006)
34. S. Kumar, V. Agrawal, R. John, B. Malhotra, Microfluidic-integrated biosensors: prospects for point-of-care diagnostics. Biotechnol. J. **8**, 1267–79 (2013)
35. C. Holland, F. Kiechle, Point-of-care molecular diagnostic systems - past, present and future. Curr. Opin. Microbiol. **8**, 504–509 (2005)
36. World Health Organization (WHO), Prevention and management of osteoporosis. Report of a WHO scientific group. WHO Technical Report Series No. 921. Geneva, Switzerland, Tech. Rep. (2003)
37. P. Khashayar, H.A. Meybodi, G. Amoabediny, B. Larijani, Biochemical markers of bone turnover and their role in osteoporosis diagnosis: a narrative review. Recent Pat. Endocr. Metab. Immune Drug Discovery **9**(2), 79–89 (2015)
38. N.A.M. Society, Management of osteoporosis in postmenopausal women: 2010 position statement of the North American menopause society. Menopause **17**(1), 25–54 (2010)
39. A. Vasudev, A. Kaushik, Y. Tomizawa, N. Norena, S. Bhansali, An LTCC-based microfluidic system for label-free, electrochemical detection of cortisol. Sens. Actuators B Chem. **182**, 139–46 (2013)
40. P. Auroux, D. Iossifidis, D. Reyes, A. Manz, Micro total analysis systems. 2. Analytical standard operations and applications. Anal. Chem. **74**(12), 2637–52 (2002)
41. V. Goral, P. Yuen, Microfluidic platforms for hepatocyte cell culture: new technologies and applications. Ann. Biomed. Eng. **40**, 1244–1254 (2012)
42. A. Bange, H.B. Halsall, W.R. Heineman, Microfluidic immunosensor systems. Biosens. Bioelectron. **20**(12), 2488–2503 (2005)
43. D. Erickson, D. Li, Integrated microfluidic devices. Anal. Chim. Acta **507**(1), 11–26 (2004)
44. W. Zimmerman, Electrochemical microfluidics. Chem. Eng. Sci. **66**(7), 1412–25 (2011)

Chapter 2
Literature Overview

Abstract This chapter provides the results of a through literature review on three main topics:

1. Biochemical markers of bone turnover and their role in osteoporosis diagnosis.
2. Bone biosensors: knowing the present and predicting the future.
3. Protein immobilization strategies for biosensing purposes

2.1 Bone Turnover Markers and Osteoporosis

2.1.1 Introduction

In the last decade, proteome analytical techniques have made a quantum leap forward and became powerful tools in addressing a broad variety of clinical problems through improving the prediction, diagnosis, and prognostication of diseases and elucidating underlying molecular pathways [1]. Accurate assessment of the biomarkers linked to these diseases is a prerequisite to achieving these goals.

2.1.2 What Are Biomarkers

The definition of biomarkers has evolved over the past decade, with WHO suggesting that "A biomarker is any substance, structure or process that can be measured in the body or its products and influence or predict the incidence of outcome or disease" [2, 3].

Qualified biomarkers are valuable tools to diagnose patients earlier in the clinical course of a disease, to predict outcomes and guide interventional approaches, to stratify participants in clinical trials, and to monitor response to therapy, and more importantly in personalized medicine. Besides providing useful information for guiding clinical decision-making, these biomarkers are increasingly linked to specific molecular pathway deregulations and/or pathogenesis that justify the application of certain therapeutic/interventional strategies [4].

P. Khashayar, *Point-of-Care Solution for Osteoporosis Management*,
DOI 10.1007/978-3-319-55053-4_2

In other words, biomarkers are key molecular, chemical, or cellular characteristics that can be objectively measured and evaluated as an indicator for a normal biological or pathogenic process such as the onset, manifestation, or progression of a disease as well as pharmacological response to a therapeutic intervention [5]. Regarding the context of use, biomarkers can be either disease- or therapy-related.

- Disease-related biomarkers are diagnostic (used to establish the disease state), prognostic (used to provide information regarding potential clinical outcome irrespective of the treatment), or predictive (used to provide information regarding potential clinical outcomes in response to specific treatments).
- Therapy-related biomarkers provide information regarding the effectiveness of a particular drug in treating or managing a disease.

The Biomarkers and Surrogate End Point Working Group, alternatively, has classified biomarkers as [6]:

- Type 0 biomarkers: Markers of natural history of the disease that correlate with clinical indices (e.g., complete blood count)
- Type I biomarkers: Markers that represent the effects of a therapeutic intervention in accordance with drug mechanism of action (e.g., decrease in urokinase-type plasminogen activator (uPA) gene expression occurring upon dasatinib treatment in prostate cancer)
- Type II biomarkers: These markers are surrogate end points as any change in their values indicates clinical benefit (e.g., decrease in blood/urine glucose level).

Concerning their origins, biomarkers are categorized as follows (Fig. 2.1):

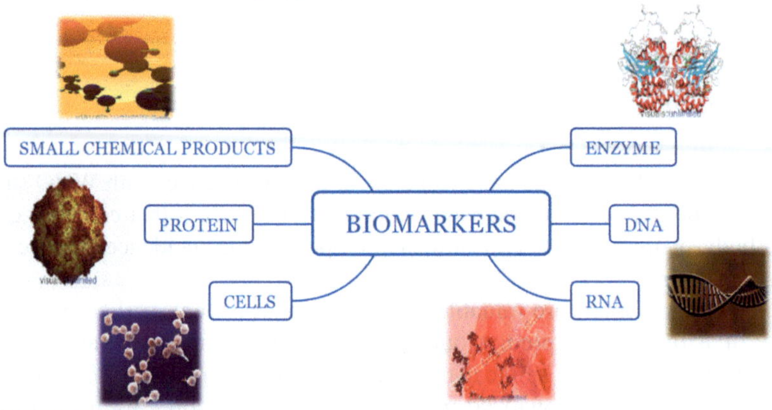

Fig. 2.1 Different origins of biomarkers

- Protein biomarkers[1]: Proteins are vital biomolecules that function as the working unit for activities, ranging from the storage and metabolism of energy to the regulation of cellular functions. An abnormal expression of proteins or expression of unique proteins is often associated with a specific disease.

  Protein biomarkers are primarily found in the blood or urine [7]. Most of them serve multiple clinical purposes during early or late disease progression. They can also be used to monitor treatment response and/or detect recurrence during follow-ups.

  Analyzing protein biomarkers involves several challenges, mainly due to their sensitivity to their ambient environment, including temperature, ionic strength, and pH. Moreover, the direct analysis of traces of proteins in crude or complex biological samples against the high background of other proteins in high abundance is difficult.
- Enzyme biomarkers
- Nucleic acids-based biomarkers (DNA and RNA)
- Small chemical products of cellular metabolism
- Direct analysis of cells.

For a biomarker to become accepted for clinical application it has to have the following characteristics:

- Being readily and consistently detectable in biological fluids, tissues, or other biological specimens.
- Being rapidly detectable and stable.
- Having high sensitivity and specificity.
- Having strong correlation with the phenotype or outcome of interest.
- Being detectable via a simple, non-invasive, and cost-effective test.
- Being consistent across genders.

2.1.3 Bone Turnover Markers (BTMs)

An important determinant of bone strength that cannot be assessed using BMD or clinical risk factor assessment tools is bone turnover [8]. BTMs, on the other hand, can be useful in assessing bone turnover through determining the number and the activity of bone cells. More specifically in terms of clinical utility, BTMs can help predict the risk of developing osteoporosis or experiencing fracture or potential response to therapy [9].

As a result, biochemical markers of bone turnover have been investigated for three main purposes [10]:

- to monitor anabolic or antiresorptive therapy in osteoporotic patients,
- to predict potential bone loss and the risk of developing osteoporosis,
- to improve assessment of fracture risk in different individuals.

[1]Considering the concept of this book, only the first category is briefly explained.

BTM measurements may also help to diagnose certain bone pathologies characterized by high bone turnover such as Paget's disease, osteomalacia, multiple myeloma, and bone metastases.

2.1.3.1 Advantages of Using BTMs

DXA is the widely available, non-invasive, and accepted method for assessing BMD with the aim of identifying individuals with osteoporosis and, possibly, those at-risk for experiencing fracture [3]. On the other hand, while BMD changes are considered as a valid end point for verifying the efficacy of osteoporosis treatments, BMD changes and fracture risk reduction rate following medication use is not measurable using DXA at early stages [4]. Moreover, as mentioned earlier in Chap. 1, DXA is not capable of assessing bone quality, and its results are influenced by advancing age and bone conditions such as osteoarthritis [5].

Furthermore, BMD alone is not efficient in diagnosing patients suffering from secondary osteoporosis, caused by certain underlying medical conditions or treatments that may interfere with the attainment of peak bone mass or may result in bone loss. The treatment process in these patients may include treating the underlying cause of the disease, which is based on additional tests including BTM assessment [11].

Moreover, since bone biopsy is invasive and cannot be repeated in clinical practice and because bone histomorphometry is less available due to the lack of specialized laboratories, the use of BTMs has gained momentum.

In this regard, BTMs have shown promising results in osteoporosis care as:

- Samples of blood and urine are easily collected
- They have a relative specificity for bone resorption or formation
- They provide complementary information to BMD
- Changes in BTMs occur earlier than that of BMD.

Figure 2.2 compares the impact of various risk factors (low BMD, a prior fracture, or high urinary CTX) on 10-year probability of hip fracture in Swedish women (women aged 65 years and over in OFELY study, and those of 80 years and over in EPIDOS study) according to age and relative risk. The difference in the impact of each of these risk factors on hip fracture probability is rather negligible. However, the combination of these risk factors improves the relative risk. In other words, the presence of two of these risk factors increases the risk of hip fracture probability by 2.0–2.6-fold, whereas using all of these risk factors, a three-fold increase is noted. Combination of these risk factors also enlarges the risk stratification range and allows the identification of individuals above a threshold risk.

BTMs reflect whole-body rates of bone resorption and formation as they provide a more representative index of the overall bone loss than that obtained by BMD values measured at specific skeletal sites containing different ratios of cancellous to cortical components with different metabolic rates [8].

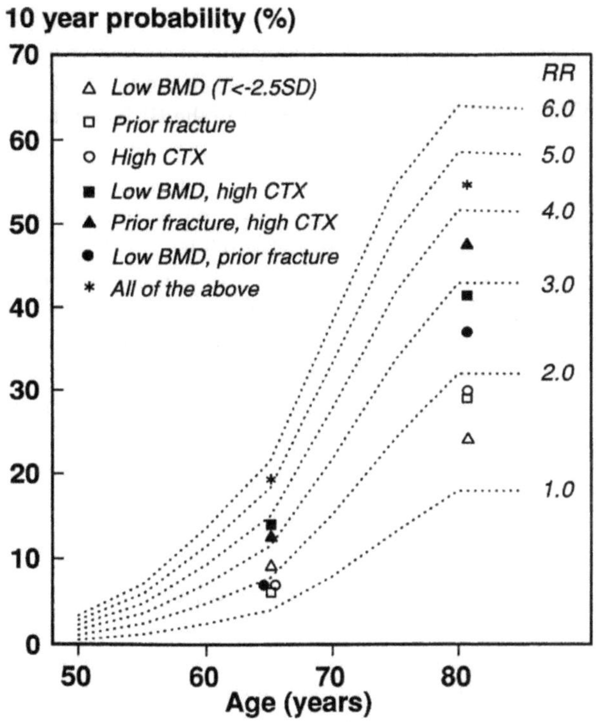

Fig. 2.2 Sensitivity of BMD and different bone markers in predicting hip fracture. "Osteoporosis International, Biochemical indices of bone turnover and the assessment of fracture probability, 13, 2002, O. Johnell. With permission of Springer"

The most effective antiresorptive treatments lower BTM values to a drug- and dose-dependent plateau within a few weeks or months (3–6 months), providing a surrogate marker with adequate sensitivity and specificity to BMD [12]. In other words, bone resorption decreases significantly after 3 weeks of antiresorptive treatment, whereas at least 6 weeks are needed for a significant decrease to be noticeable in the bone formation markers' values. At the same time considering BMD's short-term precision error (1± 1.5 %), a 2-year period is needed before at least a 3 ± 5 % change, suggestive of an acceptable response to a potent bisphosphonate, is noted on the new BMD [13].

In view of the rapid and considerable changes noted in BTM values during the course of osteoporosis treatment, these markers are also used to assess compliance and treatment response [14]. In other words, the better the compliance, the greater would be the decrease noted in BTM values, and the lower would be the fracture incidence.

BTMs can also help identify patients who benefit more from antiresorptive drugs [15]. Studies have shown that women with BTM values of at least two standard deviations higher than the mean are at a 75–80% risk of experiencing rapid bone loss [16].

Baseline BTM levels can predict treatment response as a higher increase in BMD values is reported in patients with higher turnover rate after the use of antiresorptive medication. Desai Meena et al. showed increasing levels of bone resorption markers [cross-linked telopeptides of type I collagen (CTX) and deoxypyridinoline (Dpd)] during the aging process, suggestive of a negative correlation between these markers and BMD [17]. The increase noted in the levels of CTX and Dpd was significantly higher than the changes reported in the BMD values at femur and spine.

Dresner-Pollak et al. reported that N-telopeptide cross-links (NTX), Oc, and serum parathyroid hormone explained 43% of the bone loss variability at hip in women [18]. Such a relation was not found in men. Some studies, however, have opposed the idea. For instance, Drake et al. suggested that it is not possible to identify osteoporotic men who would be good candidates for alendronate therapy based on biochemical or hormonal markers [19]. In another study, per contra, fracture risk was reported to be higher in men with increased P1NP or beta CTX levels but not TRAP5b [20].

Several studies have shown that biochemical markers of bone remodeling are independent predictors of fracture, especially in spine and hip [15]. This is mainly explained by the fact that high rate of bone loss is a risk factor for fracture, independent of BMD values.

2.1.3.2 Disadvantage of Using BTMs

Recent guidelines have embraced various roles for BTMs; they, however, have revealed differing opinions regarding the utility of BTMs in the diagnosis and management of osteoporosis [21]. Despite all, the routine use of BTMs in the clinical practice is not recommended as BTM values depend on not only bone turnover rate, but also other factors such as age and gender. Furthermore, the association between BTMs and the bone loss rate largely relies on the accuracy of the measurement techniques [22]. Analytical variability, to be discussed later on in this section, is also an important limitation to routine use of BTMs in clinical practice.

2.1.4 BTM Classification

In order to provide a list of existing BTMs, and their pros and cons in assessing bone loss, more than 950 scientific communications/articles/reports were reviewed. In brief, the articles were gathered in the following steps:

- Designing the search protocol
- Designing a quality assessment form
- Designing the data extraction form
- Searching electronic databases
- Hand searching
- Data extraction
- Data imputation and analysis.

2.1.4.1 Data Sources

International databases, including PubMed/MEDLINE, Web of Science, and Embase as well as congress abstracts conference proceedings, theses, and reports in related databases, were assessed for related Mesh terms including all subheadings:

("Osteoporosis" [MeSH] OR "Bone" [MeSH]) AND ("Biological Markers" [MeSH]).

2.1.4.2 Inclusion Criteria

All studies conducted between January 1985 and December 2011 on the accuracy of bone turnover markers in predicting osteoporosis and its complications as well as those on new methods to study BTMs for osteoporosis-related research were included. There was no restriction on the language of the publication.

2.1.4.3 Exclusion Criteria

Studies conducted on children and patients with secondary osteoporosis [chronic renal failure (CRF)] were excluded.

2.1.4.4 Main BTMs

As mentioned earlier, BTMs are classified into two main groups:

- Bone Resorption Markers:

  Most bone resorption markers, including urinary total pyridinoline (PYD), urinary free deoxypyridinoline (Dpd), urinary collagen type I cross-linked N-telopeptide (NTX), urinary or serum collagen type I cross-linked C-telopeptide (CTX), are the products of degradation of bone collagen. Non-collagenous proteins such as bone sialoprotein or tartrate-resistant acid phosphatase (TRAP) are also in this group. Increased bone resorption induces microarchitectural deterioration of bone tissue, negatively affecting bone strength.
- Bone Formation Markers:

  Bone formation markers such as bone-specific alkaline phosphatase, Oc, and serum type I procollagen (C-terminal/N-terminal) can express different phases of osteoblast development and function [23].

A list of the most important BTMs and a short description on them is provided in the following. More detailed information in this regard could be found in the article entitled "Biochemical markers of bone turnover and their role in osteoporosis diagnosis: A narrative review," published in "Recent Patents on Endocrine Metabolic & Immune Drug Discovery" by the author and her team in 2015 (Table 2.1) [24].

Table 2.1 Existing bone turnover markers

	Marker	Source	Clinical interpretation
Resorption	CTX	Degradation of bone matrix	It can significantly predict the risk of hip fracture, independent of bone mass [26]. They are also useful in monitoring bone metabolic activity and thus can be used to observe the treatment process in post-menopausal women [27]. Elevated urinary levels of native and age-related forms of CTX (alphaL/alphaD CTX ratio) are reported in patients with Paget's disease or osteoporosis. Postmenopausal women with low alpha/beta ratio of urinary CTX excretion are at an increased risk of experiencing fracture [28]
	NTX	Degradation of bone matrix	Compared with serum levels, urinary NTX is more sensitive and efficient in assessing bone resorption rate during osteoporosis treatment [29]. N-telopeptide levels could be used not only as a tool for osteoporosis screening but also for selecting patients with inflammatory bowel disease at-risk of developing osteoporosis [30]
	Hyp	Degradation of bone matrix (NTX, CTX, and ICTP are catabolized to smaller molecules such as Dpd, galactosyl hydroxylysine (GHYL), hydroxyproline (Hyp) [31])	Hyp constitutes 12±14% of the total amino acid content of mature collagen. It had long been used as the only marker of bone resorption; this is while significant amounts of urinary Hyp derive from the degradation of the newly synthesized collagen in tissues other than bone and mainly from the diet [32]. Only 10% of Hyp released during the bone resorption process is excreted in the urine
	Dpd/Pyd	Degradation of bone matrix (GHYL, the integral parts of bone collagen, are released into the circulation during the degradation of mature collagen)	Compared with Hyp, Pyd and Dpd have a better accuracy and discrimination power in distinguishing postmenopausal osteoporotic women. The measurement of these two components is not influenced by the degradation of newly synthesized collagen nor diet. They follow a circadian rhythm and are higher in early morning Pyd levels are relatively higher in females and individuals with vertebral deformities. [33] Pyd levels have a negative correlation with BMD at lumbar spine and could help differentiate between pre and postmenopausal women Dpd is a useful marker for the diagnosis of osteoporosis and other metabolic bone diseases [34]. While considered as a more specific bone marker than Pyd, Dpd can help predict the risk of vertebral fractures. Urinary excretion of Dpd is positively correlated with histomorphometric parameters of bone resorption

	TRAP	Osteoclast (The iron-containing enzyme is synthesized by osteoclasts and alveolar macrophages [35]. Its production is not linked specifically to osteoclast-mediated collagen degradation, and can be secreted by all active osteoclasts)	Enzymes originating from erythrocytes and platelets as well as many other circulating inhibitors affect total TRAP levels [36]. TRACP5b, however, is a unique indicator of the number of active osteoclasts in the body and, thus a useful indicator of bone resorption. Increased serum levels of TRAP5b are reported in certain diseases such as primary or secondary hyperparathyroidism, Paget's and metastatic bone disease where bone resorption and the number of active osteoclasts are increased [36]
	CatK	Osteoclast	Serum levels of CatK may be useful in the diagnosis of chronic bone disorders such as osteoporosis. Considering the absence of any correlation between CatK and other biomarkers including CTX or BAP, some studies suggest that serum levels of CatK are not suitable for identifying osteoporotic women [37]
	Sclerostin	Osteocyte	Decreased serum sclerostin levels are reported in patients with primary hyperparathyroidism, indicating that sclerostin is down regulated by PTH [38]. Several variables including age and serum levels of hormones such as parathyroid, estradiol, and follicle-stimulating hormone (FSH) influence serum levels of sclerostin in premenopausal women. The Wnt antagonist plays an important role in postmenopausal osteoporosis [39]
	OPG	Osteoclast	Unlike sRANKL, there is a significant positive association between OPG levels and BMD values at lumbar spine in postmenopausal women, suggesting that OPG is a good predictor of osteoporosis in elderly women [40]
Formation	Oc	Osteoblasts and odontoblasts (with a small amount being released during matrix resorption)	Oc is the most abundant non-collagenous protein in bone, comprising 1–2% of total protein [41]. The biosynthesis of Oc is regulated by several hormones and growth factors; vitamin K is responsible for its carboxylation. Cathepsin K and matrix metalloproteinases break Oc into smaller fragments (undercarboxylated osteocalcin (ucOc) and serum N-terminal osteocalcin (N-MID Oc)) during bone resorption. These fragments are rapidly cleared by the kidneys

(continued)

Table 2.1 (continued)

	Marker	Source	Clinical interpretation
	BALP	Osteoblast	BALP has regulatory role in the mineralization of the osteoids, and its measurement is helpful in diagnosis of bone loss in primary (postmenopausal and senile osteoporosis) and secondary osteoporosis (due to metabolic diseases, renal osteodystrophy, Paget's disease, etc.) as well as in monitoring antiresorptive therapy [42]
	P1NP	Newly synthesized bone matrix	P1NP is the most useful marker in this group and women with high baseline P1NP levels are reported to respond better to bisphosphonate treatment [43]
	P1CP	Newly synthesized bone matrix	Different studies have shown a significant correlation between serum P1CP levels and bone formation rate [44]. Unlike P1NP, the clinical application of P1CP in the assessment of metabolic bone diseases is still not clear
	BSP	Osteoblasts and odontoblasts	BSP accounts for 5–10% of the non-collagenous matrix of the bone. Its levels decrease rapidly following intravenous bisphosphonate treatment [45]

CTX: Cross-linked telopeptides of type 1 collagen (carboxyterminal telopeptide), NTX: N-terminal ross-linking telopeptides of type I collagen (amino terminal telopeptide), Dpd: Deoxypyridinoline, Pyd: Pyridinoline, TRAP: Tartrate-resistant acid phosphatase, CatK: Cathepsin K, Hyp: Hydroxyproline, BSP: Bone sialoprotein, OPG: Osteoprotegerin, Oc: Osteocalcin, BALP: Bone-specific alkaline phosphatase, P1NP: Procollagen type 1 N-terminal propeptide, P1CP: Procollagen type 1 C-terminal propeptide

The IOF, the International Federation of Clinical Chemistry and Laboratory Medicine (IFCC), and National Bone Health Alliance (NBHA) formed a committee of experts in 2011 to examine the efficacy of BTMs and to make clear recommendations on their use in clinical practice [25]. They recommended CTX and P1NP as the reference analytes for studying osteoporosis in clinical practice.

2.1.5 Biomarker Setbacks (with Focus on BTMs)

The inevitable biological challenges for biomarker detection have encouraged researchers to continuously devote great effort in this field. As for osteoporosis, these challenges include:

2.1.5.1 Marker Limitations

Serum or Urine Serum-based assays are better, because of their lower imprecision and day-to-day variation (within person reliability), better reproducibility (variable ionic strength of urine samples and need for creatinine (Cr) excretion correction).

Resorption or Formation Markers Changes in resorption marker values occur earlier than that of formation markers.

Single or Multiple Markers Osteoporosis is a multifactorial disease and a single biomarker has only a limited ability to predict bone loss and fracture risk with high specificity and sensitivity [46]. Moreover, to avoid false positives, arising from population variations in the expression of a single biomarker, simultaneous evaluation of a panel of biomarkers is typically important and required.

In addition, no biomarker has been reported as an ideal screening tool that can meet the diagnostic, prognostic, and predictive requirements simultaneously. A model developed by Cosman et al. showed that urine hydroxyproline (Hyp), serum CTX, serum bone alkaline phosphatase (BALP), and calcium intake predict 42% of BMD changes at lumbar spine in untreated postmenopausal women [17]. Together, the measurement of pyridinium cross-links and BALP or Oc is also reported as an index of bone turnover rate, which can help identify fast bone losers at high risk of osteoporosis or those in need of aggressive therapy [18].

2.1.5.2 Measurement Technique Limitations

Similar to other analytes, BTMs have their specific technical and analytical limitations. The relatively poor quality assurance of the measurement techniques along with their large coefficient of variation (C.V.) and assay variability are other factors limiting the use of BTMs in clinical practice. Thus assay sensitivity and specificity need to be improved, and the techniques must be standardized and validated. Thus, the validation of new techniques to measure biomarkers for efficient osteoporosis diagnosis is quite necessary and challenging. In addition, many questions have been raised regarding how new tests will be developed, evaluated, and integrated into clinical practices.

The main limitations in BTM measurements are categorized in two main groups:

Analytical Variability Immunoassays, particularly, ELISA, have been extensively used in routine clinical diagnostics and are still the gold standard for the detection of proteins, such as BTMs, in physiological samples [47].

ELISA measurements, however, are best suited when only a few biomarkers are needed to be verified/validated on a large number of samples, as they could also be a very tedious, expensive, and time-consuming process. ELISA requires antibodies against each targeted protein/peptide and thus it is possible that high quality assays would be unavailable for the proteins of interest [48] (Fig. 2.3).

The intra-individual reproducibility of BTMs is still a concern, especially because therapeutic approaches are chosen based on a single measurement. Studies have shown an intra-individual variability of around 10% and 15–25% for serum and urine markers, respectively, suggesting that a least significant change (LSC) of at least 25–30% and 40–70% is needed before a significant change is reported for each of these markers [49].

Recent improvements in technology, particularly the introduction of automated immunoassays, have reduced the analytical C.V. of BTMs to about 5–15% [50]. However, lack of a uniform standardization technique makes the comparison of the values obtained by dissimilar methods in different laboratories difficult.

Continuous efforts have been devoted to further optimize these standard methods or to find new and better alternatives for the measurement of biomarkers. The new techniques are often fluorescence or chemiluminescence based.

More recent studies thus have focused on developing newer technologies to overcome these setbacks. Next Sect. 2.2 would provide a detailed explanation of the developed bone biosensors. Although robust and highly efficient, most of these methods still suffer from lack of accuracy, sensitivity, and specificity for clinical diagnostic applications.

Pre-analytical Variability The BTMs' pre-analytical variability should also be taken into account during their clinical interpretation [51]. This variability is broadly divided into two categories (Table 2.2):

- Uncontrollable factors such as age, gender, renal function, menopausal status, underlying disease, or recent fracture. The use of appropriate reference ranges or suitable adjustments can help overcome this variability to some extent.

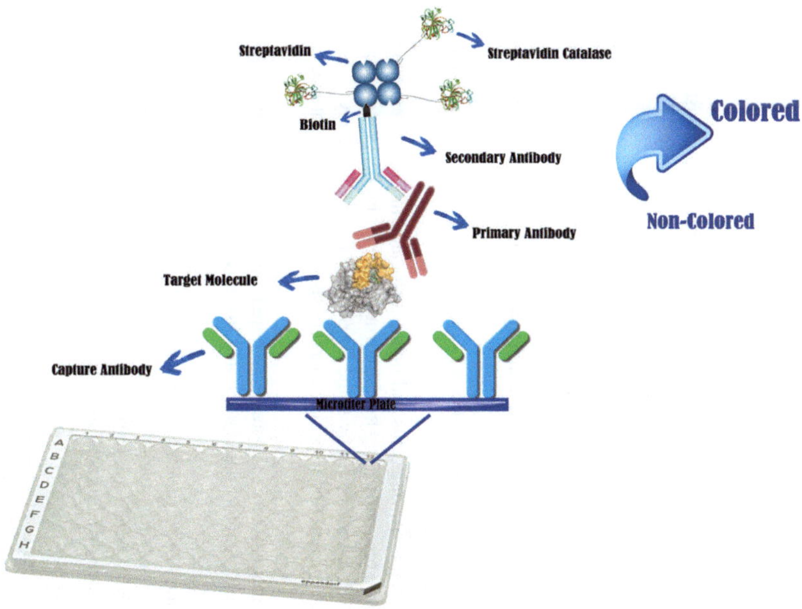

Fig. 2.3 Schematic view of the sandwich ELISA format

- Controllable factors such as circadian rhythm, menstrual, or physical activity status. The effects of these factors can be minimized by standardizing the timing and conditions under which the samples are collected.

2.1.6 Conclusion

Considering the above-mentioned facts, it is considered that the combination of several biomarkers and BMD at the same time could be useful in improving the identification of individuals at high risk for fracture [52].

As the existing BTMs could also be classified into two groups, as following, the selection of biomarkers should be made based on the final goal of the measurement system.

- BTMs that predict bone loss
  - Bone formation markers: P1NP, BALP, Oc
  - Bone resorption markers: Dpd, CTX, NTX.

Table 2.2 Sources of biological variability

Source	Nature of effect
Uncontrollable sources	
Age	BTMs are significantly higher in children, particularly in the first year of life and during puberty [53]. Its levels start to decrease after mid-puberty, and reach the baseline values in the fourth decade of life. Thereafter, the majority of BTMs remain constant in men. In women, however, a marked increase is noted at the menopause
Menopausal status	BTMs increase within a few months after the last menstrual period
Gender	Compared with their female counterparts, BTM levels are higher in men in their third and fourth decades. BTMs are higher in older women than older men [54]
Ethnicity	Bone resorption markers are lower in black youngsters compared with their white counterparts. This difference becomes apparent in women after menopause [54]
Fracture	BTM levels vary throughout the fracture healing process [55]. During the first 4 weeks of fracture, bone resorption and formation markers both increase by 20±50% and remain elevated for 6 months to 1 year
Pregnancy and lactation	During pregnancy and lactation, the maternal skeleton provides the growing fetus and infant with calcium [56]. Thus, bone turnover ratio (Dpd/Oc) rises during pregnancy; highest levels are reported during the third trimester, even higher postpartum. At term, bone resorption (NTX) and formation markers (P1NP) are increased by some 200% and 60%, respectively. After delivery, urinary NTX and CTX levels decrease but the levels of less bone-specific markers such as Pyd still continue to increase
Drug	BTMs may be decreased (glucocorticoids) or increased (anticonvulsants). Antiresorptive treatments for osteoporosis and other metabolic bone diseases such as HRT, selective estrogen receptor modulators, and bisphosphonates rapidly reduce BTMs by up to 70% [57]
Underlying diseases	Certain metabolic bone diseases such as Paget's disease, primary hyperparathyroidism, thyrotoxicosis, and acromegaly are responsible for increased levels of both bone formation and resorption markers [58]. This is while Cushing's disease or multiple myeloma may cause an increase in bone resorption and a decrease in bone formation markers. Mild decrease in bone formation and resorption markers is reported in patients with diseases such as hypoparathyroidism or hypopituitarism
Immobility	Bone formation markers decrease and resorption markers increase
Geography	Minor changes amongst countries, usually explained by differences in lifestyle
Renal function	In the elderly, an increase is noted in the levels of markers metabolized and/or excreted by the kidney (Oc and pyridinoline cross-links) due to gradual renal impairment, which occurs during aging

Between-day variability	The day-to-day variation in BTMs ranges between 16–26% for the urinary excretion of Pyd and Dpd, 7–25% for free pyridinoline, 13–35% for NTX, 12–35% for CTX, and 10–12% for TRAP [59]
Controllable sources	
Circadian rhythm	Most striking for bone resorption markers; highest values in second half of night and on waking; lowest values in afternoon and evening. Controlling the time of sample collection therefore is of great importance
Fasting status	Feeding results in a decrease in BTM; for example, s-CTX decreases by 20% after breakfast
Exercise	Changes occur but depend on type of exercise and age of subjects [60]. Moderate intensity resistance training is reported to reduce bone resorption in short time, improving the overall bone turnover rate
Menstrual status	A 10–20% variability is noted in BTM levels across the menstrual cycle [61]. Several studies have shown higher bone formation markers during the luteal phase, and higher bone resorption markers during the mid-follicular, late-follicular, and early luteal phases
Seasonal effects	While overall seasonal changes are reported to be responsible for only up to 12% of the BTM variability, some studies consider the fluctuations of vitamin D repletion as the reason behind the higher difference between BTM levels measured in summer and winter [62]
Diet	Small reduction in BTM is noted immediately following calcium supplementation. Except for HYP and CTX, diet, particularly collagen ingestion has no effect on serum/urinary levels of most BTMs [63]. The effect of the fasting-eating pattern on BTMs is, however, stronger than that of the physical activity. Fasting greatly affects the urinary and serum levels of CTX [64]

- BTMs that predict fracture
  - Bone formation markers: BALP, P1CP, P1NP, Oc
  - Bone resorption markers: TRAP, CTX, NTX.

2.2 Bone Biosensors: Knowing the Present and Predicting the Future

2.2.1 Current Approaches to Assess Bone Remodeling

As mentioned in Chap. 1, bone health depends on how the cellular mechanisms in the body create balance in the remodeling process [65]. Over the past few decades, research has been conducted in the field of biosensors to study bone-cell mechanosensation and detect the concentration of BTMs in biological samples for robust diagnosis [66].

Technological developments have enhanced assay performance through producing reliable, rapid, non-invasive, and cost-effective biosensors with improved sensitivity and specificity. While these devices are believed to help improve PoC through real-time and remote health monitoring; their use is still limited to research because of their limitations and setbacks. In the following, the latest advancements in the field of bone biosensing technologies are provided. More information in this regard could be found in an article entitled "Bone biosensors: knowing the present and predicting the future," published in the "Journal of Micromechanics and Microengineering" by the author and her team in 2016 [67].

Based on the existing literature, bone biosensors can be classified into three main categories (Table 2.3):

- Biomechanical Sensors Currently used X-ray imaging systems cannot provide accurate information about the biomechanical quality of bone regeneration at the bone graft or implant site. As a result, different sensors were developed to non-invasively assess bone growth and strength [68–71].

  Strain gauges are the gold standard for in-vivo measurement of bone strain; they are not, however, frequently used in humans because of the need for surgery to implant the sensors and cyanoacrylate adhesives to bond the sensors on the bones [72].

  Vibrational analysis of bones has also been used to assess the diseased and fractured bones in research; their clinical application, however, is limited due to lack of reliability and reproducibility of the results [73, 74].

Table 2.3 Existing bone biosensors

Device	Characteristics	Pros and cons
IMPACT3500 [85]	*Technology:* Strain gauge	
	Measures: Strain (Elastic range, fatigue limit, occurrence of fractures	*Pros:* Real-time study
	Usage: Monitoring implant strain and deformation of femoral implants	*Cons:* Invasive, non-biodegradable, dependency of sensor response substantially on the position of the neutral axis
	LoD: Elastic range: 300 μs (100-400 μs), fatigue limit: 150 μs	
Biomechanical bone sensor [76, 77]	*Technology:* Piezoelectric transducers	
	Measures: Frequency response function (FRF)	*Pros:* Direct evaluation
	Usage: Monitor mechanical and structural dynamic parameters in healing bones after surgery	*Cons:* Invasive, non-biodegradable
	LoD: Damping ratio: 0.035	
Multi-axial bone sensor	*Technology:* Piezoresistive shear stress sensor	
	Measures: Stress, resolution, response to tensile, and bending loads	*Pros:* Direct evaluation
	Usage: Monitor healing process at fracture/graft site	*Cons:* Invasive, non-biodegradable
	LoD: Sensitivity: 0.13 $\frac{mV}{(mA.MPa)}$ for 1.4 N full scales shear force range, mean hysteresis. Error: 3.5% [86]. Resolution to 100 Pa, in 1 s. Response to tensile and bending loads up to 250 kPa [78]	

(continued)

Table 2.3 (continued)

Device	Characteristics	Pros and cons
Sirivisoot biosensor [87]	*Technology:* CNT	
	Measures: Changes in CNT conductivity, osteoblast adhesion	*Pros:* Biodegradable, improved bone growth
	Usage: Release bone growth factor—Monitoring bone remodeling and regrowth near the implant	*Cons:* Invasive
	LoD: 20% increase in osteoblast adhesion	
Nanofet sensor [88]	*Technology:* Label-free nanowire field effect transistor	
	Measures: Proteins released by osteocytes under mechanical stress	*Pros:* Non-invasive, label-free
	Usage: Studying load-induced remodeling process and amount of bone formed around the implant	*Cons:* Indirect evaluation
	Location: Serum	
	LoD: 10 fg/mL	
Automated assay	*Technology:* Automated assay [89]	
	Measures: P1NP	*Pros:* Fully automated
	Usage: Assessing serum P1NP levels (useful in osteoporosis)	*Cons:* Single marker
	LoD: 5 μg/L	
Multiplex automated assay	*Technology:* Automated assay	
	Measures: P1NP, CTX, Oc, PTH [90]	*Pros:* Multiple markers, fully automated
	Usage: Assessing serum levels of several biomarkers (useful in osteoporosis)	*Cons:* Trained staff, laboratory equipment
	LoD: 0.26 g/L, 0.002 g/L, 0.51 g/L, 0.39 ng/L	

Multiplex automated assay	*Technology:* Automated assay	*Pros:* Multiple markers, fully automated
	Measures: CTX, Oc, PTH [91]	*Cons:* Trained staff, laboratory equipment
	Usage: Assessing serum levels of several biomarkers (useful in osteoporosis)	
Calcium microchip [92]	*Technology:* Reflectance measurements from arsenazo III	
	Measures: Calcium ions	*Pros:* Non-invasive
	Usage: Assessing serum calcium levels (useful in different diseases)	*Cons:* Single marker
	LoD: 2.68e-5 M	
TRAP sensor [93]	*Technology:* Electrochemical cantilever	*Pros:* Non-invasive, high accuracy, low blood sample
	Measures: TRAP, TRAP 5a, TRAP 5b	*Cons:* Single marker
	Usage: Assessing serum TRAP levels (useful in osteoporosis)	
CTX sensor [83]	*Technology:* Label-free electrochemical	
	Measures: CTX	*Pros:* Non-invasive, label free, high accuracy, low blood sample
	Usage: Assessing serum CTX levels (useful in osteoporosis)	*Cons:* Single marker
	LoD: 50 ng/mL	

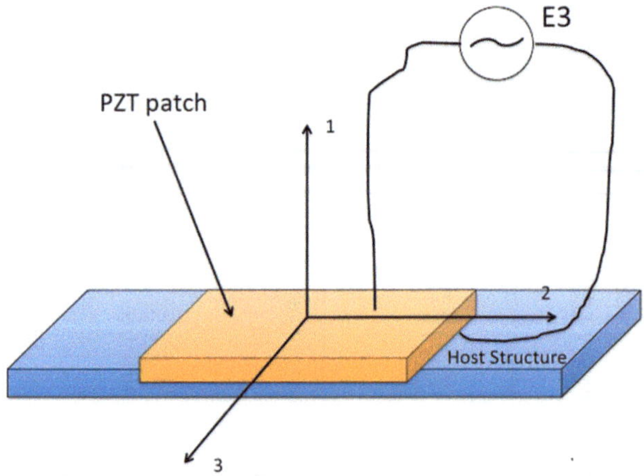

Fig. 2.4 PZT patch bonded to a 1D slender structure. "Bhalla, S. and Bajaj, S. (2008), Bone Characterization using Piezotransducers as Biomedical Sensors. Strain, 44: 475–478. Reprinted with permission from Wiley"

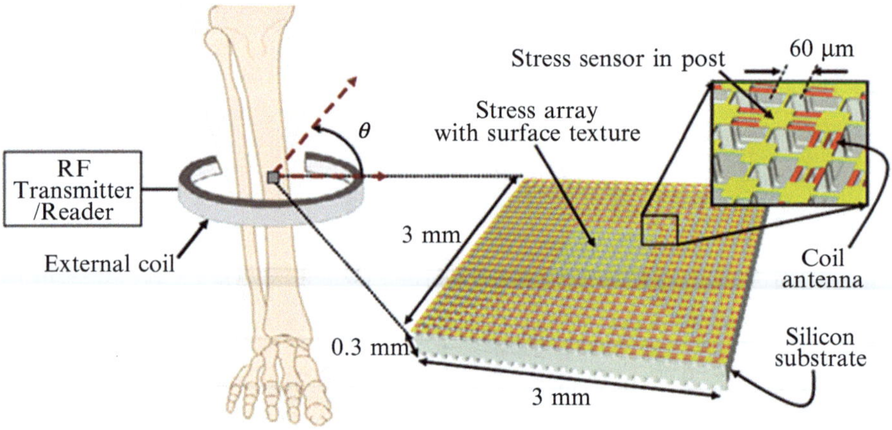

Fig. 2.5 Envisioned implantable CMOS-MEMS multiaxis stress sensor. "Fernando Alfaro et al., 2009 J. Micromech. Microeng. 19 085016. ©IOP Publishing. Reproduced with permission. All rights reserved"

Mechanical pressure sensor that could be embedded within orthopedic implants is also developed to monitor the bone healing process at the fracture site [75]. Piezoelectric sensors provide reproducible response regarding the changes in the mechanical properties of bones [76] (Fig. 2.4).

These sensors are also used to monitor capsule formation or new cell growth through studying the changes in mechanical properties of the tissue surrounding the implant [77, 78] (Fig. 2.5).

Later on the biodegradable versions of such devices were developed to overcome the need for a second surgery to remove the implanted device from the healed fracture site.

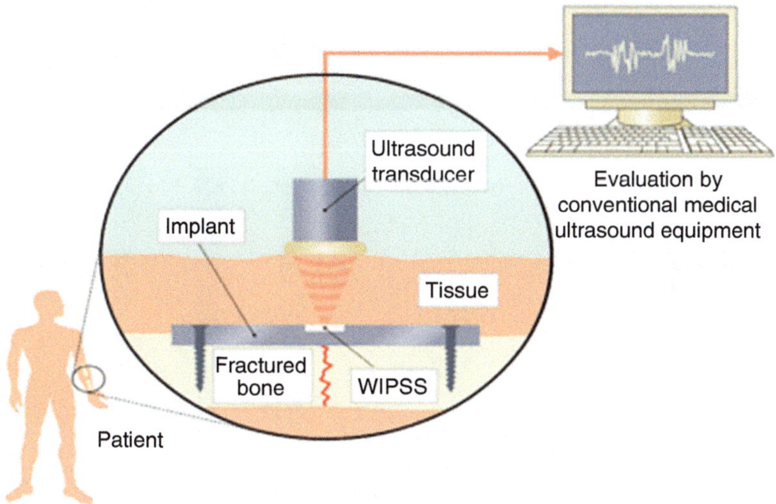

Fig. 2.6 Schematic of the WIPSS sensor concept. "F Umbrecht et al. 2010 J. Micromech. Microeng. 20 085005. ©IOP Publishing. Reproduced with permission. All rights reserved"

A biocompatible wireless implantable passive strain sensor (WIPSS) was also developed for monitoring the deformation of orthopedic implants based on the hydro-mechanical amplification effects [79] (Fig. 2.6).

- Multiplex Automated Assays While the biomechanical sensors provide information on pressure and strain levels on live bones, measuring BTM levels describes the status of the cellular and extracellular components of the skeletal matrix, reflecting bone metabolism. When interpreted correctly, these inexpensive and comparatively non-invasive tools can help diagnose and treat metabolic bone diseases [80].

  Compared with traditional techniques, multiplex protein arrays could be miniaturized and detect multiple analytes simultaneously with higher sensitivity [81].
- Label-free biosensors Label-free biosensors do not require labeling and thus allow complexes to be screened with minimal assay development [82]. They are not routinely used in clinical practice as most of them were designed to assess a single marker, which is not sufficient for assessing bone health [24, 32].

  In an attempt, Yun et al. developed an electrochemical label-free immunosensor for quantitative measurement of CTX [83] (Fig. 2.7).

2.2.2 Conclusion and Future Directions

This section sought to provide an insight into the current state of the art in the bone assessment field, stressing that new strategies are needed in clinical practice. As

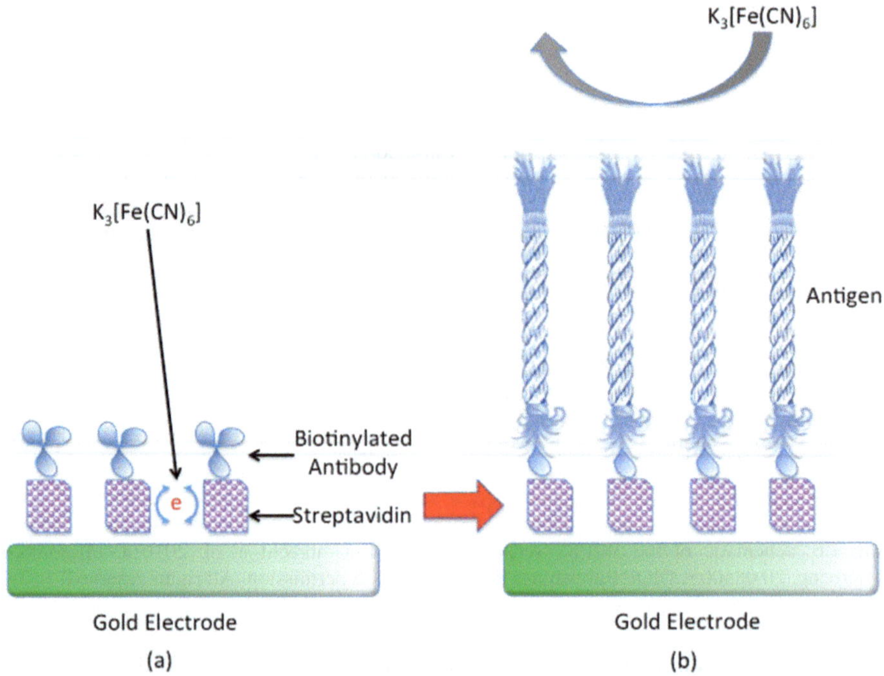

Fig. 2.7 Schematic representation of a label-free immunosensor for bone maker detection. (**a**) A self-assembled monolayer of dithiodipropionic acid deposited on a gold surface with streptavidin immobilized next as a self-assembled monolayer. Then the biotinylated antibody was bound to the streptavidin. (**b**) Antigen–antibody binding reaction and how it hinders the interfacial electron transfer reaction of $K_3[Fe(CN)_6]$[2]

mentioned earlier, BMD and other non-invasive imaging methods tell little about changes in bone material composition and biomechanical design during the healing process [65]. In order to overcome these setbacks, biomechanical sensors (group 1) were developed to provide more information about shear stress and biomechanical quality of bone regeneration at the graft or implant site [84].

These sensors could also help with monitoring the biomechanics of the fractured bone and the remodeling process of the live bone after the surgery outpatient visits [94, 95]. However, there are still concerns about their biodegradability over time. It could be possible that in the near future nano-sized biodegradable sensors would be implanted on bones to monitor their mechanical properties, utilizing advanced wireless telemetry to further facilitate physicians' decision-making process.

[2]This figure is adapted from Yun et al. [83].

Sensors classified under the second and third group gather information on the metabolic status of the bone through measuring serum or urine levels of different BTMs. Multiplex Automated Assays (group 2) can assess several markers at the same time but require expensive equipments and skilled technicians. Label-free biosensors (group 3) do not need skilled technicians but can only assess a single marker at a time [50].

2.3 Protein Immobilization Strategies for Biosensing Purposes

Protein immobilization strategies can be classified based on their selectivity, ranging from non-specific binding with low control on protein orientation (random adsorption), to more complex methods with high controllability, reproducibility, and stability (site-specific functionalization) [96]. An overview is provided in Table 2.4.

2.3.1 Physisorption

The most simple and inexpensive approach to immobilizing proteins on a surface is physisorption (physical adsorption), where proteins can be adsorbed to various surfaces via non-covalent interactions/intermolecular forces (such as van der Waals, hydrogen bonding interactions, or combination of both) [97]. Although physical adsorption offers some advantages (simplicity, no need for toxic reagents or sophisticated chemical protocols and possible reusability of the substrate), it generally leads to a surface coated with proteins in a random orientation or via multiple binding sites, which can give rise to many disadvantages (reducing the accessibility to its functional sites or conformational changes influencing the proteins' activity) [98, 99].

An additional drawback is that physisorption is generally weak, and thus the adsorbed layer is not stable and highly dependent on environmental conditions such as pH, ionic strength, and temperature. Immobilization density depends on protein size, as well as physicochemical surface properties; thus spacers such as polyethylene glycol (PEG) are widely used to reduce steric hindrance particularly when the immobilization density is too high [100].

2.3.2 Electrostatic Interaction

Electrostatic or ionic interactions provide a stronger binding of immobilized molecules and thus are more frequently exploited in biochemical assays [101]. Ionic interactions generally occur between the negatively charged gold surface and the positively charged terminals on the protein. Typical positively charged functional

Table 2.4 Most common immobilization techniques, their pros and cons

Technique		Pros	Cons
Physisorption		Inexpensive; simple; no need for toxic reagents or sophisticated chemical protocols; possible reusability of the substrate	Random orientation of the antibody or multiple binding sites, reduced accessibility to the functional sites or conformational changes influencing protein activity, weak attachment
Electrostatic interaction		Simplicity, reversibility, stronger binding, and less conformational change compared with physisorption	Limited use because of need for highly charged support and suitable condition for antibody/enzyme to remain both strongly bound and fully active
Covalent binding	NHS EDC-amine Carboxylate-EDC+NHS-amine Carboxyl chemistry Thiol chemistry Tyrosine and tryptophan	Stable nature	Reduced activity of proteins (because of linkage on active sites), need for toxic reagents, long incubation times, complicated chemistry
Bioaffinity immobilization	Avidin-biotin Protein A/G-Ab Site-specific, covalent immobilization Affinity tags	High selectivity, high specificity, oriented	High cost, complicated protocols protein immobilization, reversibility

groups such as protonated amine (R-NH_3^+) and quaternary ammonium cations (NR_4^+) as well as negatively charged carboxylate (R-COO^-) and sulfonate (R-SO_3^-) ones are mainly involved in the electrostatic interactions.

While this method does not change protein conformation to a substantial extent, the usage of highly charged supports, which can interact with charged substrates or products, can limit its use [102]. Moreover, despite its simplicity and reversibility, it is difficult to find suitable conditions under which the enzyme/antibody remains both strongly bound and fully active.

2.3.3 *Covalent Binding*

Immobilization of proteins based on the formation of covalent bonds is among the most widely used techniques, mainly because of the stable nature of the formed bonds [103]. However, covalent linkage may suffer from reduced activity of proteins (by forming linkage on active sites), toxic reagents, long incubation times, and complicated chemistry [104].

The activation processes are generally designed to generate electrophilic groups on the support, which in the coupling step, react with the strong nucleophiles from the proteins. In other words, for covalent binding, the immobilization surface is activated via reactive reagents and then reacts with amino acid residues on the protein, not involved in the reaction mechanism, and forms an irreversible linkage. The most frequently used reactions involve lysine (ϵ-amino group), cysteine (thiol group), and aspartic and glutamic acids (carboxylic acid group) as the side chains of the amino acids. Bifunctional spacer molecules are commonly used in this approach.

2.3.3.1 Amine Chemistry

The amino groups of proteins (-NH_2) are the most commonly used moieties for covalent immobilization.

N-Hydroxysuccinimide (NHS)

NHS is the most commonly used agent that reacts with protein amine groups, with the formation of stable peptide bonds [105, 106]. NHS is also often used as an activating reagent for carboxylic acids. Activated acids (basically esters with a good leaving group) can react with amines to form amides [107] (Fig. 2.8).

1-Ethyl-3-(3-Dimethylamonipropyl) Carbodiimide (EDC)-Amide

EDC, the most frequently used carbodiimide, reacts with carboxylic acid to form an o-acrylisourea intermediate, which subsequently reacts with primary amines to form

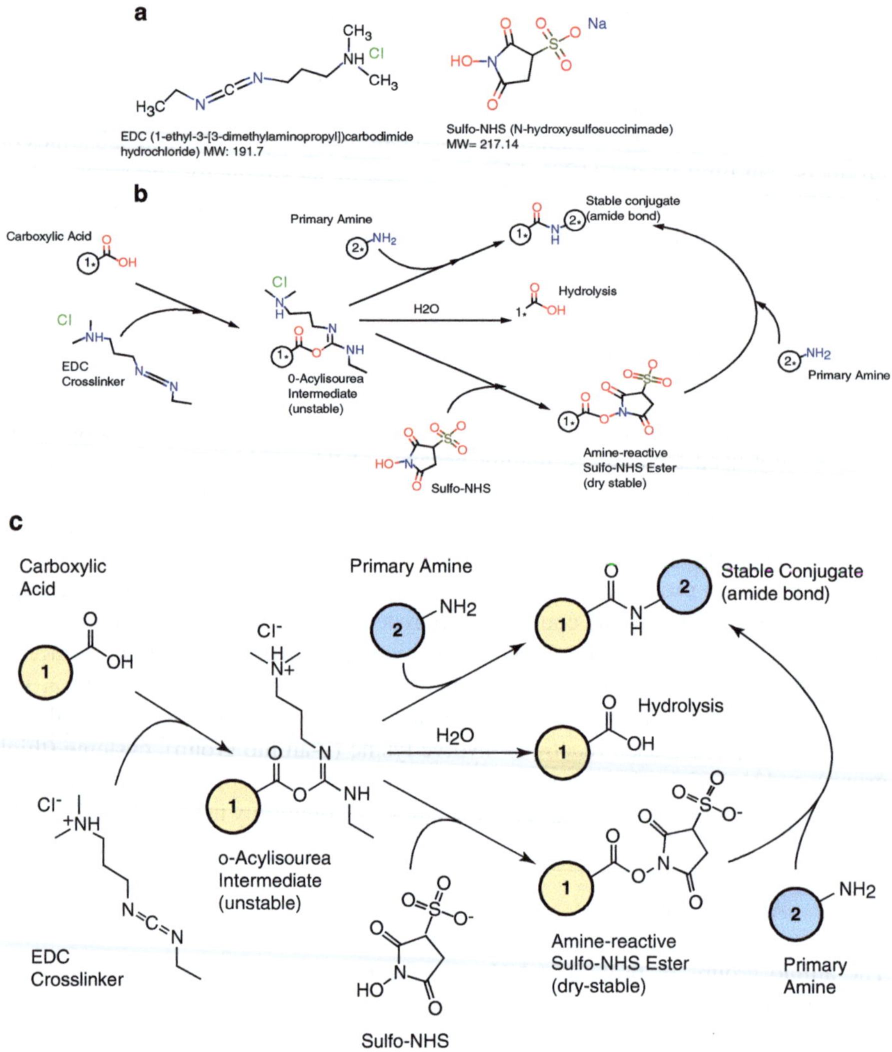

Fig. 2.8 Chemical structure of (**a**) EDC and (**b**) sulfo-NHS, (**c**) EDC and sulfo-NHS cross-linking reaction scheme. Molecules (1) and (2) can be peptides, proteins, or any chemicals that have respective carboxylate and primary amine groups[3]

amide bonds [108, 109]. Reaction of o-acrylisourea with amines is slow and can be hydrolyzed in aqueous solution. Thus, EDC is usually used with NHS (Fig. 2.8).

Carboxylate-EDC+NHS-Amine

NHS esters can be formed using carbodiimide (EDC), NHS, and carboxylates to immobilize proteins [110]. The carboxylic group would later be activated with the NHS ester, and then covalently linked to the amine groups of antibodies (Fig. 2.8).

Other chemistries used to immobilize proteins via the exposed amine groups include sulfhydryl (epoxide), iso(thio)cyanate, azlactone, p-nitrophenyl ester, glutaraldehyde, sulfonyl chlorides, and cyanogen bromide [111].

2.3.3.2 Carboxyl Chemistry

The immobilization of proteins using the carboxylic side chains is of great interest, as the amino acids glutamic and aspartic acid constitute a major fraction of the surface exposed amino acids [112]. The carboxylic acid function of these amino acids, along with the C-terminus, can react with amines using the routine coupling chemistry also used for solid phase peptide synthesis. This coupling reaction is activated by a carbodiimide like N,N'-dicyclohexyl carbodiimide (DCC) or EDC and results in a rapid and quantitative formation of a peptide bond.

2.3.3.3 Thiol Chemistry

Free cysteines have a relatively low natural abundance in proteins and thus are an interesting tag for immobilization reactions as random orientations or multiple contact points are less a problem [113]. They contain a reactive thiol (SH) group, which is more nucleophilic than a primary amino group. In proteins with no free cysteines, the amino acids can be inserted at a site of interest by site-directed mutagenesis. Maleimides and vinyl-sulfone groups are the most commonly used chemical groups to selectively modify surfaces towards covalent protein coupling via the free thiols [114].

A covalent bond can also be formed between substrates functionalized with disulfide probes via a thiol-disulfide exchange under oxidative conditions. This reaction is selective for cysteines, but similar to natural disulfide bonds, these couplings are not resistant to reducing agents like mercaptoethanol or dithiothreitol (DTT) [115].

2.3.3.4 Tyrosine and Tryptophan

Tyrosine and tryptophan are relatively the rarest amino acids on protein surfaces and can be genetically introduced without changing the overall charge state or redox sensitivity [116]. They, however, are often overrepresented near active sites of proteins; thus targeting them for modification or immobilization needs careful consideration and specific conditions (acidic environment), which may have undesirable affects on the protein's structure and function [117].

2.3.4 Bioaffinity Immobilization

The bioaffinity interaction or biospecific adsorption is based on the principle of complementary biomolecules interactions, which represents its biggest advantages, high selectivity, high specificity, and oriented protein immobilization [118]. Additionally, bioaffinity immobilization can be reversed using chemical treatment, pH change, or heat treatment [119]. Avidin-biotin, protein A/G-antibody, genetically engineered protein affinity ligands, DNA hybridization, and aptamers are common approaches in this strategy.

2.3.4.1 Avidin-Biotin

One of the well-established affinity-based immobilization techniques is based on the strong interaction between (strept)avidin (66–69 kDa tetrameric glycoprotein) and biotin (water-soluble vitamin B), and biotinylated proteins [120]. Avidin binds to biotin via an exceptionally strong non-covalent interaction, which is rapid and nearly insensitive to pH, temperature, proteolysis, and denaturing agents. It is also stable, showing no significant loss of activity over time. One downside of using the avidin-biotin system is the high cost of the proteinaceous-binding reagent.

2.3.4.2 Protein A/G-Antibody

Protein A and G are both popular antibody [Immunoglobulin G (IgG)] immobilizing reagents, extracted from bacteria. They specifically bind to the constant fragment crystallizable region (Fc region) of IgG. Thus, the variable fragment antigen-binding (Fab) region of IgG is accessible to Ag binding [121]. An additional benefit of using protein A/G is that antibody can be detached by acid treatment and the surface made reusable.

2.3.4.3 Site-Specific Covalent Immobilization

Site-specific covalent immobilization, an oriented covalent coupling of a protein to a substrate, is performed in a controllable way via a unique chemical group or sequence introduced in the protein at a site-specific location [122]. Several bioorthogonal chemistries, also known as "click" reactions such as copper or ring-strain catalyzed azide-alkyne cyclo-additions, Staudinger-ligations, Diels–Alder cyclo-additions, thiol-ene additions, and oxime formation, are commonly used in this regard.

2.3.4.4 Affinity Tags

One of the most important advantages of fusion proteins (chimeric proteins), also known as affinity tags, is the possibility to attach a generating recombinant fusion part, the preparation of which can sometime be very expensive or complicated for immobilization of various proteins [123]. Polyhistidine tags, in which the C- or N-terminus of proteins are genetically engineered to have an oligohistidine (His) segment that specifically reacts with metal ions immobilized on the matrix, are of the most common applications of this technique [124].

Although the affinity of these tags is much weaker than that of the streptavidin-biotin linkage, the immobilization could be reversed by adding competing chelating agents such as ethylenediaminetetraacetic acid (EDTA), and the controlled orientation of immobilized proteins is another benefit. The interference of large number of metal-binding proteins or additional interactions between the immobilized protein and affinity support, on the other hand, is a serious limitation.

2.3.5 Conclusion

As described above, each immobilization mechanism has its drawbacks and advantages with relative importance depending on fabrication constraints and application requirements. Thus, attempts have been made to address the limitations of individual mechanisms, aiming for better protein activity, more control over protein orientation, less steric hindrance, more protein patterning density, and less non-specific protein adsorption using a combined method.

References

1. A. Patel, E. Puzas, J. Baumhauer, Practical osteoporosis management: topical review. Foot Ankle Int. **31**(4), 354–360 (2010)
2. A. Morrison, T. Fan, S. Sen, L. Weisenfluh, Epidemiology of falls and osteoporotic fractures: a systematic review. Clin. Outcomes Res. **5**, 9–18 (2013)

3. K. Briot, C. Roux, What is the role of dxa, qus and bone markers in fracture prediction, treatment allocation and monitoring? Best Pract. Res. Clin. Rheumatol. **19**(6), 951–964 (2005)
4. S.L. Bonnick, L. Shulman, Monitoring osteoporosis therapy: bone mineral density, bone turnover markers, or both? Am. J. Med. **119**(4) Suppl 1, S25–31 (2006)
5. C.A. Cefalu, Is bone mineral density predictive of fracture risk reduction? Curr. Med. Res. Opin. **20**(3), 341–349 (2004)
6. M. Buyse, D. Sargent, A. Grothey, A. Matheson, A. de Gramont, Biomarkers and surrogate end points—the challenge of statistical validation. Nat. Rev. Clin. Oncol. **7**, 309–317 (2010)
7. H. Johansson, J.A. Kanis, A. Oden, O. Johnell, E. McCloskey, BMD, clinical risk factors and their combination for hip fracture prevention. Osteoporos. Int. **20**(10), 1675–1682 (2009)
8. R.K. McCormick, Osteoporosis: integrating biomarkers and other diagnostic correlates into the management of bone fragility. Altern. Med. Rev. **12**(2), 113–145 (2007)
9. T. Bhattarai, K. Bhattacharya, P. Chaudhuri, P. Sengupta, Correlation of common biochemical markers for bone turnover, serum calcium, and alkaline phosphatase in post-menopausal women. Malays. J. Med. Sci. **21**(1), 58–61 (2014)
10. J. Lenora, K.K. Ivaska, P. Gerdhem, Use of bone turnover markers in osteoporosis. Clin. Rev. Bone Miner. Metab. **8**(1), 1–14 (2010)
11. K.D. Harper, T.J. Weber, Secondary osteoporosis. Diagnostic considerations. Endocrinol. Metab. Clin. North Am. **27**(2), 325–348 (1998)
12. D.J. Baylink, The diagnosis and management of osteoporosis. Z. Rheumatol. **59** Suppl 1, 42–44 (2000)
13. N. Yoshimura, World Health Organization, Absolute risk for fracture and who guideline. bone turnover markers as predictive factors for osteoporotic fractures. Clin. Calcium **17**(7), 1049–1057 (2007)
14. J.A. Clowes, N.F.A. Peel, R. Eastell, The impact of monitoring on adherence and persistence with antiresorptive treatment for postmenopausal osteoporosis: a randomized controlled trial. J. Clin. Endocrinol. Metab. **89**(3), 1117–1123 (2004)
15. P. Garnero, E. Hausherr, M.C. Chapuy, C. Marcelli, H. Grandjean, C. Muller, C. Cormier, G. Bréart, P.J. Meunier, P.D. Delmas, Markers of bone resorption predict hip fracture in elderly women: the epidos prospective study. J. Bone Miner. Res. **11**(10), 1531–1538 (1996)
16. K.J.L. Bell, A. Hayen, L. Irwig, M.C. Hochberg, K.E. Ensrud, S.R. Cummings, D.C. Bauer, The potential value of monitoring bone turnover markers among women on alendronate. J. Bone Miner. Res. **27**(1), 195–201 (2012)
17. P.D. Meena, M. Khatkhatay, K.B. Prakash, L. Savardekar, R. Shah, Z. Ansari, Hormonal profiles and biochemical indices of bone turnover in Indian women. Osteoporos. Int. **18**(7), 923–929 (2007)
18. R. Dresner-Pollak, R. Parker, M. Poku, J. Thompson, M. Seibel, S. Greenspan, Biochemical markers of bone turnover reflect femoral bone loss in elderly women. Calcif. Tissue Int. **59**(5), 328–333 (1996)
19. W.M. Drake, D.L. Kendler, C.J. Rosen, E.S. Orwoll, An investigation of the predictors of bone mineral density and response to therapy with alendronate in osteoporotic men. J. Clin. Endocrinol. Metab. **88**(12), 5759–5765 (2003)
20. N.B. Watts, Clinical utility of biochemical markers of bone remodeling. Clin. Chem. **45**(8), 1359–1368 (1999)
21. P. Glendenning, Markers of bone turnover for the prediction of fracture risk and monitoring of osteoporosis treatment: a need for international reference standards. Clin. Biochem. Rev. **32**(1), 45–47 (2011)
22. G. Wheater, M. Elshahaly, S.P. Tuck, H.K. Datta, J.M. van Laar, The clinical utility of bone marker measurements in osteoporosis. J. Transl. Med. **11**, 201 (2013)
23. P. Garnero, P.D. Delmas, *Biochemical Markers of Bone turnover in Osteoporosis*, vol. 2, ed. by J. Kelsey, R. Marcus, D. Feldman (Academic, New York, 2001), pp. 459–477
24. P. Khashayar, H.A. Meybodi, G. Amoabediny, B. Larijani, Biochemical markers of bone turnover and their role in osteoporosis diagnosis: a narrative review. Recent Pat. Endocr. Metab. Immune Drug Discovery **9**(2), 79–89 (2015)

25. D.C. Bauer, P. Garnero, S.L. Harrison, J.A. Cauley, R. Eastell, K.E. Ensrud, E. Orwoll, for the Osteoporotic Fractures in Men (MrOS) Research Group, Biochemical markers of bone turnover, hip bone loss, and fracture in older men: the MrOS study. J. Bone Miner. Res. **24**(12), 2032–2038 (2009)
26. M.J. Seibel, H. Woitge, C. Scheidt-Nave, G. Leidig-Bruckner, A. Duncan, P. Nicol, R. Ziegler, S.P. Robins, Urinary hydroxypyridinium crosslinks of collagen in population-based screening for overt vertebral osteoporosis: results of a pilot study. J. Bone Miner. Res. **9**(9), 1433–1440 (1994)
27. C. Dane, B. Dane, A. Cetin, M. Erginbas, Effect of risedronate on biochemical marker of bone resorption in postmenopausal women with osteoporosis or osteopenia. Gynecol. Endocrinol. **24**(4), 207–213 (2008)
28. H. Hoshino, M. Takahashi, K. Kushida, T. Ohishi, T. Inoue, The relationships between the degree of beta-isomerization of type i collagen degradation products in the urine and aging, menopause and osteoporosis with fractures. Osteoporos. Int. **9**(5), 405–409 (1999)
29. Y. Abe, H. Ishikawa, A. Fukao, Higher efficacy of urinary bone resorption marker measurements in assessing response to treatment for osteoporosis in postmenopausal women. Tohoku J. Exp. Med. **214**(1), 51–59 (2008)
30. N. Bregenzer, P. Erban, H. Albrich, G. Schmitz, S. Feuerbach, J. Schölmerich, T. Andus, Screening for osteoporosis in patients with inflammatory bowel disease by using urinary n-telopeptides. Eur. J. Gastroenterol. Hepatol. **14**(6), 599–605 (2002)
31. O.S. Donescu, M.C. Battié, T. Videman, J. Risteli, D. Eyre, The predictive role of bone turnover markers for bmd in middle-aged men. Aging Male **9**(2), 97–102 (2006)
32. M.J. Seibel, Biochemical markers of bone turnover: part i: biochemistry and variability. Clin. Biochem. Rev. **26**(4), 97–122 (2005)
33. A. Behre, J. Janott, M. Pfohl, H. Schatz, A. Pfeiffer, Clinical value of urinary pyridinium crosslinks as osteoporosis markers: evaluation of a population survey of vertebral osteoporosis. Med. Klin. (Munich) **96**(7), 378–382 (2001)
34. P. Bettica, L. Moro, S.P. Robins, A.K. Taylor, J. Talbot, F.R. Singer, D.J. Baylink, Bone-resorption markers galactosyl hydroxylysine, pyridinium crosslinks, and hydroxyproline compared. Clin. Chem. **38**(11), 2313–2318 (1992)
35. J.M. Halleen, S.L. Alatalo, A.J. Janckila, H.W. Woitge, M.J. Seibel, H.K. Väänänen, Serum tartrate-resistant acid phosphatase 5b is a specific and sensitive marker of bone resorption. Clin. Chem. **47**(3), 597–600 (2001)
36. B. Habermann, C. Eberhardt, M. Feld, L. Zichner, A.A. Kurth, Tartrate-resistant acid phosphatase 5b (trap 5b) as a marker of osteoclast activity in the early phase after cementless total hip replacement. Acta Orthop. **78**(2), 221–225 (2007)
37. D. Adolf, T. Wex, O. Jahn, C. Riebau, W. Halangk, S. Klose, S. Westphal, H. Amthauer, S. Winckler, S. Piatek, Serum cathepsin k levels are not suitable to differentiate women with chronic bone disorders such as osteopenia and osteoporosis from healthy pre- and postmenopausal women. Maturitas **71**(2), 169–172 (2012)
38. M.-S. Ardawi, A. Al-Sibiany, T. Bakhsh, A. Rouzi, M. Qari, Decreased serum sclerostin levels in patients with primary hyperparathyroidism: a cross-sectional and a longitudinal study. Osteoporos. Int. **23**(6), 1789–1797 (2012)
39. X.-J. Xu, L. Shen, Y.-P. Yang, F.-R. Lu, R. Zhu, B. Shuai, C.-G. Li, M.-X. Wu, Serum sclerostin levels associated with lumbar spine bone mineral density and bone turnover markers in patients with postmenopausal osteoporosis. Chin. Med. J. (Engl.) **126**(13), 2480–2484 (2013)
40. Y. Chiba, T. Onouchi, T. Ikeda, J. Adachi, Y. Tamura, T. Horiuchi, Implications of measuring soluble receptor activators of nuclear factor-kappab ligand and osteoprotegerin in bone metabolism of elderly women. Gerontology **55**(3), 275–280 (2009)
41. C.M. Gundberg, J.B. Lian, S.L. Booth, Vitamin k-dependent carboxylation of osteocalcin: friend or foe? Adv. Nutr. **3**(2), 149–157 (2012)
42. J. Ambroszkiewicz, J. Gajewska, T. Laskowska-Klita, Bone alkaline phosphatase: characteristic and its clinical applications. Med. Wieku Rozwoj. **6**(2), 99–110 (2002)

43. D.C. Bauer, P. Garnero, M.C. Hochberg, A. Santora, P. Delmas, S.K. Ewing, D.M. Black, Fracture Intervention Research Group, Pretreatment levels of bone turnover and the antifracture efficacy of alendronate: the fracture intervention trial. J. Bone Miner. Res. **21**(2), 292–299 (2006)
44. E.F. Eriksen, P. Charles, F. Melsen, L. Mosekilde, L. Risteli, J. Risteli, Serum markers of type i collagen formation and degradation in metabolic bone disease: correlation with bone histomorphometry. J. Bone Miner. Res. **8**(2), 127–132 (1993)
45. J.K. Chen, H.S. Shapiro, J.L. Wrana, S. Reimers, J.N. Heersche, J. Sodek, Localization of bone sialoprotein (bsp) expression to sites of mineralized tissue formation in fetal rat tissues by in situ hybridization. Matrix **11**(2), 133–143 (1991)
46. F. Cosman, J. Nieves, C. Wilkinson, D. Schnering, V. Shen, R. Lindsay, Bone density change and biochemical indices of skeletal turnover. Calcif. Tissue Int. **58**(4), 236–243 (1996)
47. M. Seibel, M. Lang, W. Geilenkeuser, Interlaboratory variation of biochemical markers of bone turnover. Clin. Chem. **47**, 1443–1450 (2001)
48. R. Lequin, Enzyme immunoassay (EIA)/enzyme-linked immunosorbent assay (ELISA). Clin. Chem. **51**(12), 2415–2418 (2005)
49. C. Popp-Snijders, P. Lips, J. Netelenbos, Intra-individual variation in bone resorption markers in urine. Ann. Clin. Biochem. **33**, 347–348 (1996)
50. S. Vasikaran, C. Cooper, R. Eastell, A. Griesmacher, H.A. Morris, T. Trenti, J.A. Kanis, International osteoporosis foundation and international federation of clinical chemistry and laboratory medicine position on bone marker standards in osteoporosis. Clin. Chem. Lab Med. **49**(8), 1271–1274 (2011)
51. R. Hannon, R. Eastell, Preanalytical variability of biochemical markers of bone turnover. Osteoporos. Int. **11**, Suppl 6, S30–S44 (2000)
52. O. Löfman, P. Magnusson, G. Toss, L. Larsson, Common biochemical markers of bone turnover predict future bone loss: a 5-year follow-up study. Clin. Chim. Acta **356**(1–2), 67–75 (2005)
53. M. Midtby, J. Magnus, R. Joakimsen, The tromso study: a population-based study on the variation in bone formation markers with age, gender, anthropometry and season in both men and women. Osteoporos. Int. **12**, 835–843 (2001)
54. Y. Henry, R. Eastell, Ethnic and gender differences in bone mineral density and bone turnover in young adults: effect of bone size. Osteoporos. Int. **11**, 512–517 (2000)
55. G. Cox, T. Einhorn, C. Tzioupis, P. Giannoudis, Bone turnover markers in fracture healing. J. Bone Joint Surg. Br. **92-B**(3), 329–334 (2010)
56. C. More, H. Bhattoa, P. Bettembuk, A. Balogh, The effects of pregnancy and lactation on hormonal status and biochemical markers of bone turnover. Eur. J. Obstet. Gynecol. Reprod. Biol. **106**(2), 209–213 (2003)
57. P.D. Delmas, P. Eastell, P. Garnero, M.J. Seibel, J. Stepan, and for the Committee of Scientific Advisors of the International Osteoporosis Foundation, The use of biochemical markers of bone turnover in osteoporosis. Osteoporos. Int. **11**, Suppl 6, S2–S17 (2000)
58. R. Civitelli, R. Armamento-Villareal, N. Napoli, Bone turnover markers: understanding their value in clinical trials and clinical practice. Osteoporos. Int. **20**(6), 843–851 (2009)
59. A. Blumsohn, R. Hannon, A. al Dehaimi, R. Eastell, Short-term intraindividual variability of markers of bone turnover in healthy adults. J. Bone Miner. Res. **9**, Suppl 1, p. S153 (1994)
60. T. Whipple, B. Le, L. Demers, V. Chinchilli, M. Petit, N. Sharkey, N. Williams, Acute effects of moderate intensity resistance exercise on bone cell activity. Int. J. Sports Med. **25**(7), 496–501 (2004)
61. A. Schlemmer, C. Hassager, J. Risteli, L. Risteli, S. Jensen, C. Christiansen, Possible variation in bone resorption during the normal menstrual cycle. Acta Endocrinol. (Copenh) **129**, 388–392 (1993)
62. H. Woitge, C. Scheidt-Nave, C. Kissling, G. Leidig-Bruckner, K. Meyer, A. Grauer, Seasonal variation of biochemical indexes of bone turnover: results of a population-based study. J. Clin. Endocrinol. Metab. **83**, 68–75 (1998)

63. M. Lang, P. Haag, H. Schmidt-Gayk, R. Ziegler, M. Seibel, Influence of ambient storage conditions and of diet on the measurement of biochemical markers of bone metabolism. Calcif. Tissue Int. **55**, 497 (1995)
64. A. Schlemmer, C. Hassager, Acute fasting diminishes the circadian rhythm of biochemical markers of bone resorption. Eur. J. Endocrinol. **140**(4), 332–337 (1999)
65. E. Seeman, P.D. Delmas, Bone quality–the material and structural basis of bone strength and fragility. N. Engl. J. Med. **354**(21), 2250–2261 (2006)
66. L. Liu, T.J. Webster, In situ sensor advancements for osteoporosis prevention diagnosis, and treatment. Curr. Osteoporos Rep. **14**(6), 386–395 (2016)
67. P. Khashayar, G. Amoabediny, B. Larijani, J. Vanfleteren, Bone biosensors: knowing the present and predicting the future. J Micromech. Microeng. (2016)
68. G. Bergmann, F. Graichen, J. Siraky, H. Jendrzynski, A. Rohlmann, Multichannel strain gauge telemetry for orthopaedic implants. J. Biomech. **21**(2), 169–176 (1988)
69. G.Y. Yang, V.J. Bailey, Y.-H. Wen, G. Lin, W.C. Tang, J.H. Keyak, Fabrication and characterization of microscale sensors for bone surface strain measurement, in *IEEE Sensors* (2004), pp. 1355–1358
70. S.J. Hoshaw, D.P. Fyhrie, Y. Takano, D. Burr, C. Milgrom, A method suitable for in vivo measurement of bone strain in humans. J. Biomech. **30**, 521–524 (1997)
71. N. Elvin, A. Elvin, M. Spector, Implantable bone strain telemetry sensing system and method, US Patent 6034296, Tech. Rep. (2000)
72. G.J. Pratt, Apparatus for establishing in vivo, bone strength, US Patent 4421119, Tech. Rep. (1983)
73. F. Gattiker, F. Umbrecht, J. Nevenschwander, U. Sennhauser, C. Hierold, Novel ultrasound read-out for a wireless implantable passive strain sensor system, in *Proc IEEE Sensors* (2008), pp. 20–23
74. F. Nogata, A. Shimamoto, T. Habu, Estimation of in-vivo bone strength using ultrasound signals. Int. J. Mod. Phys. B **17**(8–9), 1381–1387 (2003)
75. Mechanical sensor teams up with ultrasound for bone monitoring (2008), Available: http://www.medgadget.com/2008/11/mechanical_sensor_teams_up_with_ultrasound_for_bone_monitoring.html
76. S. Bhalla, S. Bajaj, Bone characterization using piezotransducers as biomedical sensors. Strain **44**(6), 475–478 (2008)
77. J.W. Bender, H.I. Friedman, V. Giurgiutiu, C. Watson, M. Fitzmaurice, M.L. Yost, The use of biomedical sensors to monitor capsule formation around soft tissue implants. Ann. Plast. Surg. **56**(1), 72–77 (2006)
78. A. Fernando, W. Lee, C. Phil, M. Mark, K.F. Gary, Design of a multi-axis implantable mems sensor for intraosseous bone stress monitoring. J. Micromech. Microeng. **19**, 8 (2009)
79. F. Umbrecht, P. Wagli, S. Dechand, F. Gattiker, J. Neuenschwander, U. Sennhauser, C. Hierold, Wireless implantable passive strain sensor: design, fabrication and characterization. J. Micromech. Microeng. **20**, 085005 (2010)
80. D.J. Leeming, P. Alexandersen, M.A. Karsdal, P. Qvist, S. Schaller, L.B. Tankó, An update on biomarkers of bone turnover and their utility in biomedical research and clinical practice. Eur. J. Clin. Pharmacol. **62**(10), 781–792 (2006)
81. R. Eastell, P. Garnero, C. Audebert, D.L. Cahall, Reference intervals of bone turnover markers in healthy premenopausal women: results from a cross-sectional European study. Bone **50**(5), 1141–1147 (2012)
82. B.E. Rapp, F.J. Gruhl, K. Länge, Biosensors with label-free detection designed for diagnostic applications. Anal. Bioanal. Chem. **398**(6), 2403–2412 (2010)
83. Y.-H. Yun, A. Bhattacharya, N.B. Watts, M.J. Schulz, A label-free electronic biosensor for detection of bone turnover markers. Sensors (Basel) **9**(10), 7957–7969 (2009)
84. A. Leonidou, M. Moazen, P. Lepetsos, S. Graham, G. Macheras, E. Tsiridis, The biomechanical effect of bone quality and fracture topography on locking plate fixation in periprosthetic femoral fractures. Injury **46**(2), 213–217 (2015)

85. F. Burny, M. Donkerwolcke, F. Moulart, R. Bourgois, R. Puers, K. Van Schuylenbergh, M. Barbosa, O. Paiva, F. Rodes, J.B. Bégueret, P. Lawes, Concept, design and fabrication of smart orthopedic implants. Med. Engl. Phys. **22**(7), 469–479 (2000)
86. M. Hsieh, Y. Fang, M. Ju, J. Ho, S. Ting, Development of a new contact-type piezoresistive micro-shear-stress sensor, in *Proc. SPIE 4755, Design, Test, Integration, and Packaging of MEMS/MOEMS* (2002), p. 285
87. S. Sirivisoot, C. Yao, X. Xiao, B.W. Sheldon, T.J. Webster, Developing biosensors for monitoring orthopedic tissue growth, in *MRS Proceedings*, vol. 950 (2006)
88. J. Li, Fabrication and characterization of nano-fet biosensors for studying osteocyte mechanotransduction. Master's thesis, Department of Mechanical and Industrial Engineering Institute of Biomaterials and Biomedical Engineering University of Toronto (2011)
89. P. Garnero, P. Vergnaud, N. Hoyle, Evaluation of a fully automated serum assay for total N-terminal propeptide of type I collagen in postmenopausal osteoporosis. Clin. Chem. **54**(1), 188–196 (2008)
90. A. Claudon, P. Vergnaud, C. Valverde, A. Mayr, U. Klause, P. Garnero, New automated multiplex assay for bone turnover markers in osteoporosis. Clin. Chem. **54**(9), 1554–1563 (2008)
91. H. Schmidt-Gayk, E. Spanuth, J. Kötting, R. Bartl, D. Felsenberg, J. Pfeilschifter, F. Raue, H. Roth, Performance evaluation of automated assays for beta-crosslaps, n-mid-osteocalcin and intact parathyroid hormone (biorose multicenter study). Clin. Chem. Lab Med. **42**(1), 90–95 (2004)
92. P. Caglar, S.A. Tuncel, N. Malcik, J.P. Landers, J.P. Ferrance, A microchip sensor for calcium determination. Anal. Bioanal. Chem. **386**(5), 1303–1312 (2006)
93. Y. Chung, Y. Liu, Biosensor and method for bone mineral density measurements, US Patent 20050059875, Tech. Rep. (2005)
94. C. Moß, N. Weinrich, K. Seide, J. Müller, Wireless recording system for long-term monitoring of bone fracture healing, in *IFMBE Proceedings*, vol. 25/7 (World Congress on Medical Physics and Biomedical Engineering, Munich, 2009), pp. 446–448
95. S. Bhalla, Bone characterization using piezo-transducers as bio-medical sensors, in *Smart Materials in Structural Health Monitoring, Control and Biomechanics* (Springer, Berlin, 2012).
96. R.E. Steen, D. Ta, D. Cortens, B. Billen, W. Guedens, P. Adriaensens, Protein engineering for directed immobilization. Bioconjug. Chem. **24**(11), 1761–1777 (2013)
97. A. Ansari, Q. Husain, Potential applications of enzymes immobilized on/in nano materials: a review. Biotechnol. Adv. **30**, 512–523 (2012)
98. K. Nakanishi, T. Sakiyama, Y. Kumada, K. Imamura, H. Imanaka, Recent advances in controlled immobilization of proteins onto the surface of the solid substrate and its possible application to proteomics. Curr. Proteomics **5**, 161 (2008)
99. C. Mateo, J. Palomo, G. Fernandez-Lorente, J. Guisan, R. Fernandez-Lafuente, Improvement of enzyme activity, stability and selectivity via immobilization techniques. Enzyme Microb. Technol. **40**, 1451–1463 (2007)
100. J. Yakovleva, R. Davidsson, A. Lobanova, M. Bengtsson, S. Eremin, T. Laurell, J. Emnéus, Microfluidic enzyme immunoassay using silicon microchip with immobilized abs and chemiluminescence detection. Anal. Chem. **74**, 2994 (2002)
101. D. Kim, A. Herr, Protein immobilization techniques for microfluidic assays. Biomicrofluidics. **7**, 4 (2013)
102. E. Benešová, B. Králová, *Affinity Interactions as a Tool for Protein Immobilization, Affinity Chromatography* (InTech, Rijeka, 2012)
103. B. Brena, P. González-Pombo, F. Batista-Viera, Immobilization of enzymes: a literature survey, in *Immobilization of Enzymes and Cells*. Methods in Molecular Biology, vol. 1051 (Springer, New York, 2013)
104. J. Křenková, F. Foret, Immobilized microfluidic enzymatic reactors. Electrophoresis **25**, 3550 (2004)

105. J. Guisán, Agarose-aldehyde gels as supports for immobilization- stabilization of enzymes. Enzyme Microb. Technol. **10**, 375–382 (1998)
106. B. Wang, C. Guo, M. Zhang, B. Park, B. Xu, High-resolution single-molecule recognition imaging of the molecular details of ricin-aptamer interaction. J. Phys. Chem. B **116**, 5316–5322 (2012)
107. Z. Pei, H. Anderson, A. Myrskog, G. Dunér, B. Ingemarsson, T. Aastrup, Optimizing immobilization on two-dimensional carboxyl surface: pH dependence of antibody orientation and antigen binding capacity. Anal. Biochem. **398**, 161–168 (2010)
108. Y. Wang, H.-H. Lai, M. Bachman, C.E. Sims, G.P. Li, N.L. Allbritton, Covalent micropatterning of poly (dimethylsiloxane) by photografting through a mask. Anal. Chem. **77**, 7539–7546 (2005)
109. D. Bartczak, A. Kanaras, Preparation of peptide-functionalized gold nanoparticles using one pot EDC/sulfo-NHS coupling. Langmuir **27**(16), 10119–10123 (2011)
110. T. Didar, A. Foudeh, M. Tabrizian, Patterning multiplex protein microarrays in a single microfluidic channel. Anal. Chem. **84**, 1012 (2012)
111. T. Boller, C. Meier, S. Menzler, EUPERGIT oxirane acrylic beads: how to make enzymes fit for biocatalysis. Org. Process Res. Dev. **6**, 509–519 (2002)
112. L. Cao, Immobilised enzymes: science or art? Curr. Opin. Chem. Biol. **9**, 217–226 (2005)
113. E. Basle, N. Joubert, M. Pucheault, Protein chemical modification on endogenous amino acids. Chem. Biol. **17**, 213–227 (2010)
114. G. Hermanson, The chemistry of reactive groups: vinylsulfone derivatives, Chap. 2.7, in *Bioconjugate Techniques*, 2nd edn. (Academic, Rockford, 2008)
115. Y. Kim, S. Ho, N. Gassman, Y. Korlann, E. Landorf, F. Collart, S. Weiss, Efficient site-specific labeling of proteins via cysteines. Bioconjug. Chem. **19**, 786–791 (2008)
116. N. Joshi, L. Whitaker, M. Francis, A three-component mannich-type reaction for selective tyrosine bioconjugation. J. Am. Chem. Soc. **126**, 15942–15943 (2004)
117. M. Gauthier, H. Klok, Peptide/protein-polymer conjugates: synthetic strategies and design concepts. Chem. Commun. **23**, 2591–2611 (2008)
118. B. Brena, F. Batista-Viera, Immobilization of enzymes, in *Immobilization of Enzymes and Cells*, 2nd edn. (Humana Press, New Jersey, 2006)
119. S. Laib, B. MacCraith, Immobilization of biomolecules on cycloolefin polymer supports. Anal. Chem. **79**, 6264 (2007)
120. K. Holland-Nell, A. Beck-Sickinger, Specifically immobilised aldo/keto reductase AKR1A1 shows a dramatic increase in activity relative to the randomly immobilised enzyme. Chembiochem **8**, 1071–1076 (2007)
121. J. Yakovleva, R. Davidsson, M. Bengtsson, T. Laurell, J. Emnéus, Microfluidic enzyme immunosensors with immobilised protein a and g using chemiluminescence detection. Biosens. Bioelectron. **19**, 21 (2003)
122. R. Blanco, J. Calvete, J. Guisán, Immobilization-stabilization of enzymes. Variables that control the intensity of the trypsin (amine)-agarose-(aldehyde) -multipoint attachment. Enzyme Microb. Technol. **11**, 353–359 (1989)
123. J. Arnau, C. Lauritzen, G.E. Petersen, J. Pedersen, Current strategies for the use of affinity tags and tag removal for the purification of recombinant proteins. Protein Expr. Purif. **48**(1), 1–13 (2006)
124. S. Knecht, D. Ricklin, A.N. Eberle, B. Ernst, Oligohis-tags: mechanisms of binding to ni2+-nta surfaces. J. Mol. Recognit. **22**(4), 270–279 (2009)

Chapter 3
Scientific Background

Abstract This chapter aims to introduce the scientific and technical principles applied in the experimental chapters of this work. Three major subjects will be discussed.

- The first subject is an overview on biosensors and its types. A special focus is dedicated to electrochemical sensors.
- The second part deals with different electrochemical techniques. The definition of these techniques will be described.
- The last section embraces main principles of microfluidic devices.

3.1 Overview on Immunosensors

3.1.1 Introduction

The concept of using immunological components as sensing agents was first described within an immunoassay for plasma insulin in human subjects [1]. The ELISA test, developed based on the high dissociation constants (K_d) with which antibodies bind to their target antigen, has long been the gold standard for comparison against all newly developed immunoassays and immunosensors [2]. Briefly, ELISAs involve the immobilization of a reactant (an antibody or antigen) onto a solid surface, with enzymes being used as markers for the presence and abundance of a specific antibody–antigen (Ab/Ag) interaction [3].

The widespread use of antibodies regardless of their high cost, variable affinity, and short shelf-life arises from their exceptional specificity and sensitivity towards their binding partners, i.e., the antigens (Ag). Immunoassays, therefore, make use of the sensitivity and specificity of the antibody–antigen interaction [4]. Several methods exist to transduce a signal created by the binding of antibody and analyte, each with associated advantages and disadvantages, leading to a wealth of research into fabrication of immunosensors and immunoassays [5].

Most clinical analyses are carried out by specialized staff in laboratories using desktop instruments, thus assuring the highest possible confidence in the obtained results. However, there are many cases in which a critical clinical analysis cannot

P. Khashayar, *Point-of-Care Solution for Osteoporosis Management*,
DOI 10.1007/978-3-319-55053-4_3

be performed in optimal conditions, due to lack of trained technicians or required facilities, as is often the case in underdeveloped or isolated areas. In those cases, biosensors can be the only option to make a trustworthy medical diagnosis (Fig. 3.1)

3.1.2 What Is a Biosensor

The first biosensor was described in 1962 by Clark and Lyons [7]. They immobilized glucose oxidase (GOD) on a semipermeable dialysis membrane on an amperometric oxygen electrode surface to quantify glucose concentration.

According to the International Union of Pure and Applied Chemistry (IUPAC), "A biosensor is a self-contained integrated device which is capable of providing specific quantitative or semi-quantitative analytical information using a biological recognition element (biochemical receptor) which is in direct spatial contact with a transducer element."

A biosensor should be clearly distinguished from a bioanalytical system, which requires additional processing steps, such as reagent addition, and from a bioprobe which is either disposable after one measurement, or unable to continuously monitor the analyte concentration [8]. In other words, biosensors permit the analysis of analytes with high specificity and sensitivity in a short time [9]. These techniques have the capability to be used in PoC systems and portable devices [10].

PoC testing is an "on site" test, detailed within a large proportion of research. However, none of the existing immunosensors offer a complete set of the necessary characteristics for a good PoC sensor, including full automation, portability, precision, accuracy and sensitivity, low cost, and ease of use [11]. Non-invasiveness is another important characteristic when considering PoC sensor design, preferably when samples such as urine, sweat, or saliva are used. Minimizing the pain associated with blood sampling is also possible by reducing needed sample volume [11].

The detection process in a biosensor is mainly based on the specific interaction between the analyte of interest and the recognition element (such as an electrode). As a result of this specific interaction, changes in one or several physicochemical properties (pH, electron transference, heat transfer, change of potential or mass, variation of optical properties, etc.) occur, which later on are detected and measured by a specific transducer.

It therefore could be concluded that a biosensor is composed of two main parts (Fig. 3.2):

- A bioreceptor or an immobilized sensitive biological element (enzyme, DNA probe, antibody) recognizing the analyte (enzyme substrate, complementary DNA, antigen).
- A transducer used to convert (bio)chemical signal resulting from the interaction of the analyte with the bioreceptor into an electronic one. The intensity of the generated signal is directly or inversely proportional to the concentration of the analyte [12].

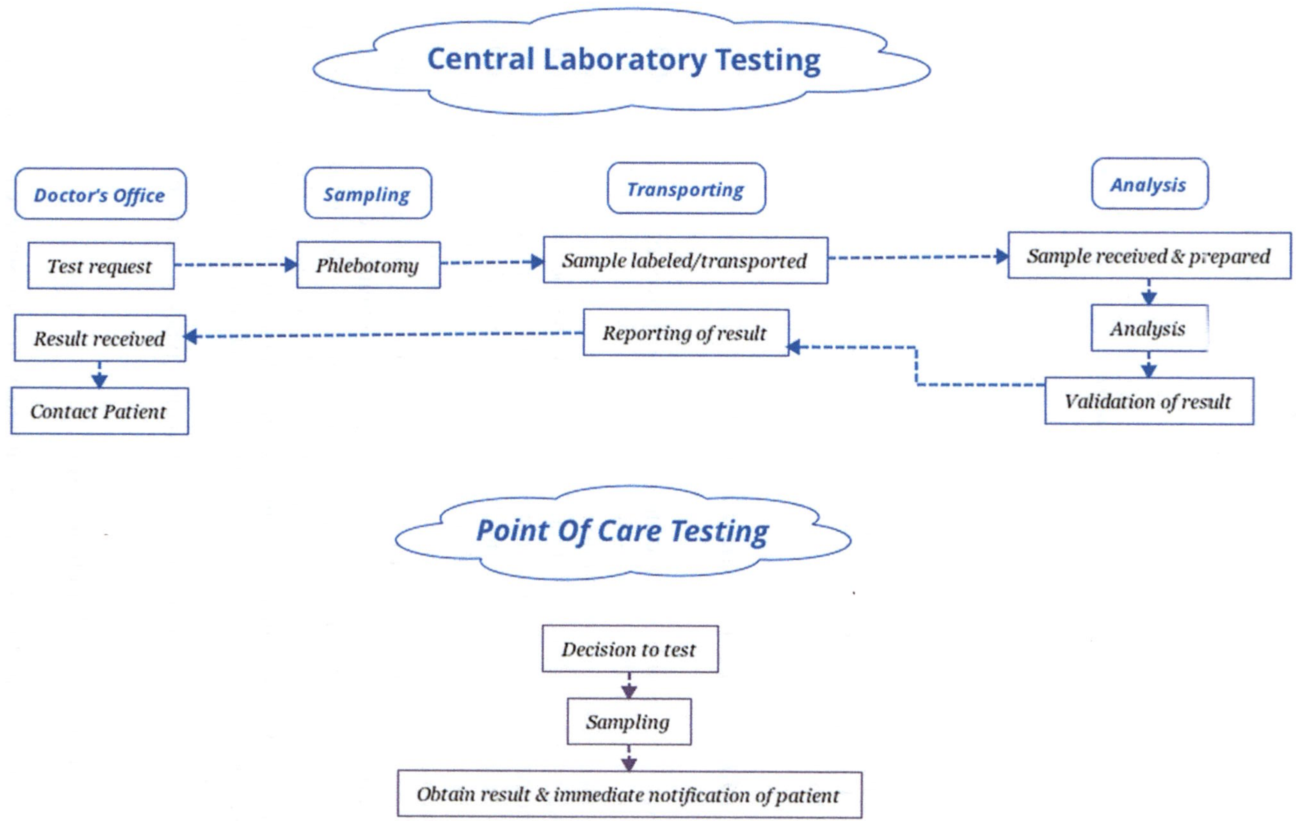

Fig. 3.1 Process of clinical testing in outpatient situations using central laboratory versus PoC methods. The processes are shown in a simplified format and may sometimes contain additional steps [6]

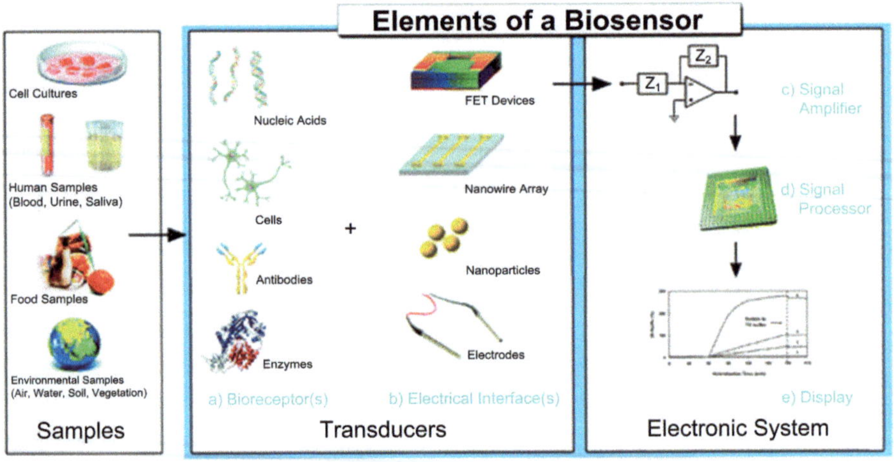

Fig. 3.2 Different types of biosensors based on their structure. "From: Grieshaber, D.; MacKenzie, R.; Voros, J.; Reimhult, E. Electrochemical Biosensors - Sensor Principles and Architectures. Sensors 2008, 8, 1400–1458"

3.1.3 Main Types of Immunoassays

Immunosensors could be divided into "direct" (label-free) or "indirect" (labeled) assays. In other words, the detection mechanism operates either directly via the Ab/Ag interaction, or an additional label (such as an enzyme or fluorescent molecule) is used to detect whether a binding event has occurred or not (Fig. 3.3). The principal challenge for many label-free protein biomarker assays is the occurrence of a relatively less significant change in the receptors' properties upon binding with the target analyte and the high specificity of the acquired responses [13].

In another classification, immunoassays are divided into "competitive" and "noncompetitive." In competitive assays, the analyte competes in proportion to its concentration with a labeled analyte for a limited number of antibody binding sites. As the concentration of the analyte increases, more labeled analyte is displaced, resulting in an increase in free labeled analytes and thus in signal. In a noncompetitive assay, such as a sandwich immunoassay, which will be discussed in more detail in the following paragraphs, the signal increases in proportion to the analyte concentration [14].

Based on another grouping, there are two main types of immunoassays.

3.1.3.1 Heterogeneous Immunoassays

In heterogeneous immunoassays, the antibody (or in some cases antigen) is immobilized on a solid support [15, 16]. The most commonly used heterogeneous immunoassay is the sandwich assay, which usually consists of a capture step with

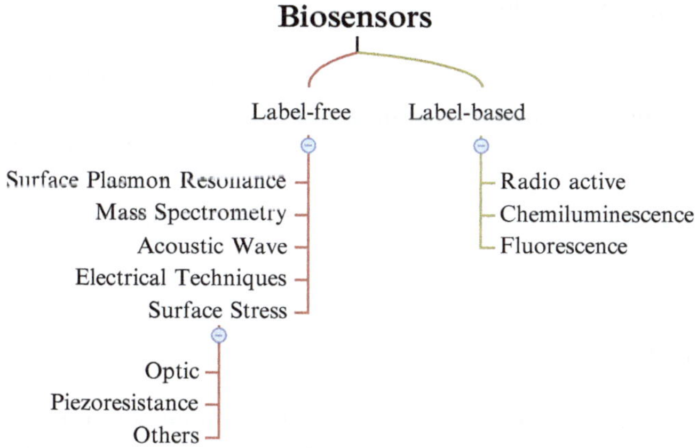

Fig. 3.3 Different types of biosensors based on the detection technique

a primary antibody followed by a second binding reaction between the captured antigen and a labeled second antibody. This is usually the preferred assay format when high sensitivity and selectivity is needed as this technique involves two separate recognition steps provided by the two antibodies that are generally selected to recognize two different epitopes on the antigen. The need for the analyte to be able to bind with two antibodies simultaneously is the main limitation, which restricts its use to molecules with a molecular weight of greater than about 1000 Da.

3.1.3.2 Homogenous Immunoassays

Homogeneous immunoassays, which rely on the formation of the antigen–antibody complex causing a change in the intensity of the label's signal, take place entirely in the solution phase. Absence of Ab immobilization and rinsing steps simplifies the assay. Absence of a solid phase that can be affected by non-specific binding is considered as an advantage for the technique [17]. Homogeneous assays are more difficult to develop and generally have poorer limits of detection and selectivity than heterogeneous assays. They are commonly used for detecting therapeutic drugs with a relatively high concentration in blood.

3.1.4 Immunosensor Components

The most commonly used components in immunosensor technology are as follows.

3.1.4.1 Carbon Nanotubes

The importance of single-walled (SWNT's) and multi-walled (MWNT's) CNTs in the development of the immunochemical devices was first reported in 1991 [18]. This was mainly due to their role in enhancing detection sensitivity (possibly due to their higher surface area, their electrocatalytic effects, lower levels of fouling, being mechanically strong and flexible, and excellent electrical and thermal conductivity) [19].

3.1.4.2 Nanoparticles

Over the past decade, more and more research has been performed on the use of nano-sized metal particles, particularly noble metal nanoparticles with controlled size, morphology, and crystal orientation, because of their different physical and chemical properties compared with the bulk material [20, 21]. The main benefit of nanoparticles is their ability to significantly enhance sensor's sensitivity and lower its detection limit, possibly through increasing the surface area onto which antibodies can be immobilized, or affecting the refraction of light in SPR-based sensors [22, 23].

Among the noble metal nanoparticles, gold nanoparticles (AuNPs) are of particular interest because of not only their stability, but also their specific physicochemical properties, including enhanced diffusion of electroactive species based on their high effective surface area, improved selectivity, catalytic activity, and higher signal-to-noise ratio and conductivity [24, 25]. Important here, AuNPs provide a stable surface for protein immobilization and overcome the need for external electron transfer mediators in electrochemical sensing [26–28]. The use of AuNPs in this work is going to be discussed later on in Chaps. 4 and 5.

3.1.5 *Detection Techniques*

As mentioned earlier, biosensors could also be categorized based on their detection techniques [29]:

- Optical [30]
  - Direct detection by monitoring light properties: fluorescence, absorbance, and luminescence/chemiluminescence.
  - Light property modulation detections: surface plasmon resonance (SPR), evanescent waves, interferometry, Raman spectrometry, fiber optics, and optical waveguides [32].

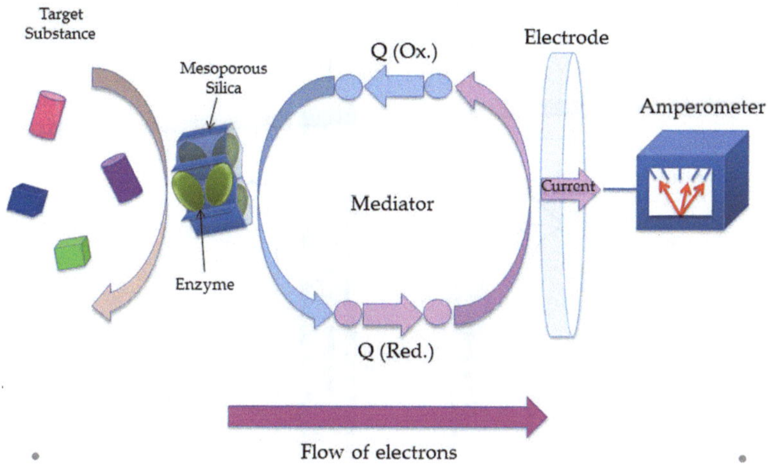

Fig. 3.4 Schematic diagram of detection mechanism in a sample enzymatic amperosensor [31]

- Non-optical
  - Electrochemical: The electrochemical detection can be used to measure the concentration of the analyte by converting its chemical signal into the electrical signal via the electrodes [33].

    Amperometry: Amperometric sensors measure the current signal generated by the redox reaction on the electrode while a constant potential is applied [34]. Amperometric detection is based on the fact that the applied voltammetric potential between the reference and working electrode would cause the oxidation or reduction of the electroactive species in the vicinity, and induce an electrical current.

    The analyte thus should be either redox active or the use of a redox active label will be necessary. The change of the current is related to the concentration of the analyte (Fig. 3.4). Such sensors consume a small percentage of analytes during measurement, which creates concentration gradients; and thus rendering responses are stir-dependent.

    The sensitivity of amperometric immunosensors can be improved through increasing the surface area onto which capture agents are immobilized and thus increasing binding events and sensitivity. This could be happen using a layer-by-layer (LBL) construction approach utilizing AuNPs or methylene blue to form a uniform and stable base [35]. Cross-linking also prevents conformational change and unfolding of the antibodies and thus markedly enhances sensitivity, longer storage times, and higher resistance to extremes of pH and temperature. The first technique is to be discussed later on in Chap. 5 and the latter here in Chap. 2.

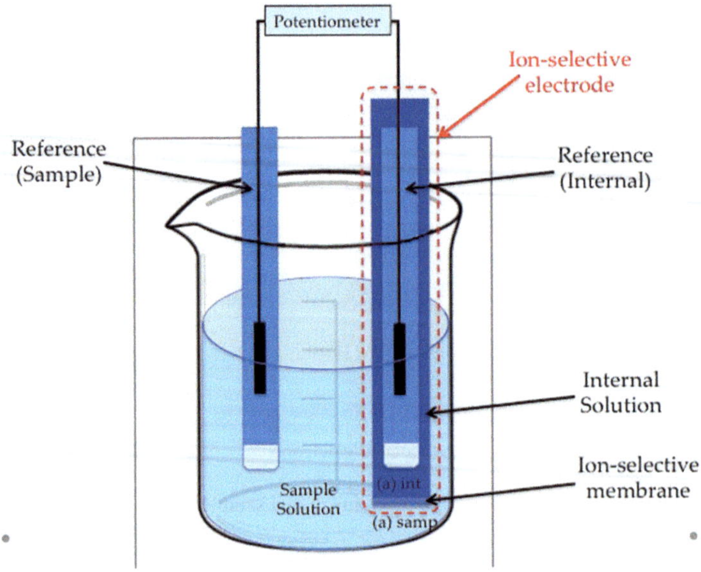

Fig. 3.5 Schematic diagram showing a typical potentiometric cell with an ion-selective electrode. The ion-selective electrode's membrane separates the sample, which contains the analyte from an internal solution containing the analyte[1]

Potentiometry: In potentiometric detection, analyte is assessed by monitoring the potential of an ion-selective electrode (usually a membrane/surface sensitive to an analyte) against a reference electrode. The reaction between the surface and the analyte generates a potential proportional to the logarithm of the electrochemically active material concentration (Fig. 3.5).

A sensor, which transduces the potential across a membrane into a digitized signal or field-effect transistors (FETs) whose semi-conductive surface potential changes upon a change in analyte concentration are examples of potentiometric sensors [36].

Conductometry: Conductometric biosensors can measure the changes of the electrical conductivity of a medium, mainly a change in the composition of the solution, between electrodes at a series of frequencies. Conventional contacted and capacitively coupled contactless conductivity detection (C4D) methods such as potential gradient detection conductometric biosensors.

Impedimetry: Impedimetric detection method is similar to the amperometric detection method but uses alternating current (AC) instead; thus the electrolysis of the electrodes can be eliminated. However, similar to the conductometric detection method, the sensitivity of the biosensor depends on the difference of the impedance between the background and the analyte.

[1]This figure is adapted from [40]

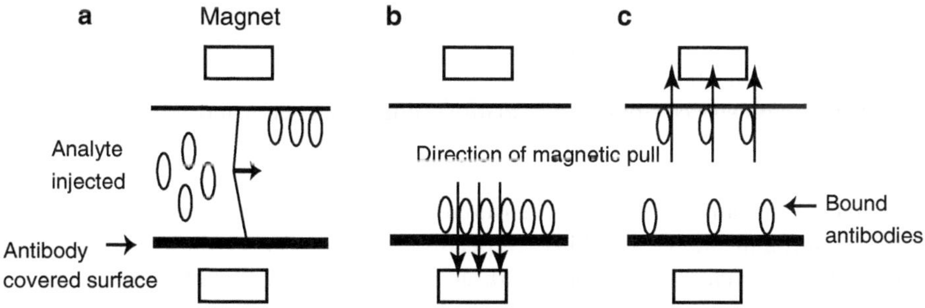

Fig. 3.6 (**a**) The microchamber is filled with analyte in solution and the antibodies are redispersed into the solution, some bind with the analyte. (**b**) Magnetic particles are attracted to the lower surface. (**c**) The free and weakly bound immunosensors which are not bound to analyte are removed from the lower surface when the upper magnet is switched on

- Magnetic immunosensors: There has been a recent surge of interest in the use of magnetic particles in biosensor development [37, 38]. Electromagnets can pull immuno-substituted beads towards a binding site and then weakly bound species can be removed by an oppositely located electromagnet. They allow a higher level of control and no extra wash stages, giving a simple one-step measurement once the sample fluid has been added [39]. In other words, they provide simplicity and rapid procurement of results that are needed for PoC (Fig. 3.6).
- Piezoelectric: Piezoelectric devices convert physical or mechanical changes into electrical energy and vice versa. Quartz crystal microbalance (QCM), which is the most common piezoelectric sensor, exploits the change in the resonance of quartz crystals upon changes in their mass. The technique allows binding of antigen to antibody (when one of these is immobilized on the crystal surface) to be measured electrically [41, 42].
- Mass spectrometry
- Other techniques such as nuclear magnetic resonance (NMR) spectroscopy, magneto-resistive, and acoustical.

3.2 Electrochemistry: Principles

One key issue to take into consideration in biosensing platform integration is the increasing demand for higher sensitivity and selectivity at minimal costs, and the possibility to monitor biosensing in real-time, especially for point-of-care platforms, for which simplicity should also be considered [43]. In view of these facts, calorimetric and electrochemical approaches are the most promising biosensing platforms due to their simplicity, high sensitivity and specificity, and no need for expensive and complex instrumentations. Moreover, such approaches allow for multiplexing biosensing enabling their application in a range from low to high throughput diagnostics.

Electrochemistry combines an oxidation–reduction chemical reaction that occurs in a liquid phase with an electrical circuit that is completed by solid conductive materials. In this way the kinetics of the reaction that occurs on the electrode (catalytic or binding events) is measured by the signal produced from electron transfer through the system [44]. Depending on the type of the analysis, either potential or current may be measured, and the amount of this electrical signal is proportional to analyte concentration [45, 46].

As mentioned earlier, electrochemistry provides powerful analytical techniques with advantages such as instrumental simplicity, moderate cost, and portability. Another advantage to electrochemical assays is that they are bioselective and highly sensitive to very low concentrations. Considering these advantages, an electrochemical biosensor was designed to fulfill the final objective of this work.

In the following, some basic electrochemistry principles are discussed.

A standard electrochemical experiment, referred to as a three-electrode setup, uses a reference electrode (REF), working electrode (WE), and counter/auxiliary electrode (AUX) (Fig. 3.7).

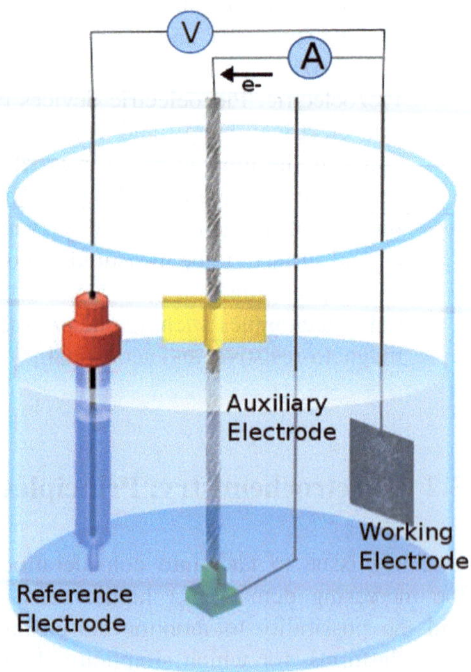

Fig. 3.7 Schematic representation of a three-electrode system. "Image courtesy of the University of Texas at Dallas. Reprinted with permission"

- Working electrode: The electrochemical reaction takes place on a polarizable electrode, which is used as the WE.
- Reference electrode: REF, which establishes the electrical potential against other potentials, supplies a constant potential standard for introducing the electrochemical reaction on the working electrode. The ideal reference electrode provides a stable, known potential so that any change in E_{cell} is attributed to the analyte's effect on the potential of the metal indicator electrode [Eq. (3.1)],[2]

$$E_{\mathrm{cell}} = E_{\mathrm{ind}} - E_{\mathrm{ref}} + E_j \tag{3.1}$$

  where E_{ind} refers to the metal indicator potential and E_j to the junction potential.

  In theory, no current passes through the reference electrode, and its potential remains constant at the value specified by the analyst. In reality, however, a drop is noted in potential between the working and reference electrodes, since they cannot be placed infinitely close to each other [44]. Common reference electrodes are "Standard Hydrogen Electrode," "Calomel Electrode," and "Silver/Silver Chloride Electrode."
- Auxiliary electrode: When a certain potential is applied to the WE with respect to the REF, the redox reaction occurs. The reaction generates a current, proportional to the concentration of the analyte, which can be measured between the WE and the AUX electrodes.

The appeal of potentiostatic techniques is that the thermodynamics of the system can be controlled by adjusting the potential. For a general electrochemical reaction, this relationship is described by the Nernst equation [47].

In other words, the Nernst equation describes the relationship between the potential applied to an electrode and the concentration of the redox species on the electrode surface [33]. If an electrode is at equilibrium with the solution in which it is immersed in, the electrode will have a potential, invariant with time, which is thermodynamically related to the composition of the solution. In solution, the "Oxidant" is capable of being reduced to the "Reductant" on the electrode by the following reversible electrochemical reaction [Eq. (3.2)]:

$$\text{Oxidant} + ne^- \leftrightarrows \text{Reductant} \tag{3.2}$$

The Nernst equation relates the potential E, which is applied to the electrode and the concentration of the Oxidant and Reductant on the electrode surface [Eqs. (3.3), (3.4)]:

$$E = E_0 - \left(\frac{RT}{nF}\right) \ln Q \tag{3.3}$$

[2]Metal indicator electrodes are paired with a reference electrode so that the cell potential will be directly influenced by the target analyte concentration in the solution.

also written as,

$$E = E_0 - \frac{0.0592V}{n}\log[Q] \tag{3.4}$$

where E is the potential difference between the electrodes (V); E_0 is the standard reduction potential (V); R is the universal gas constant ($R = 8.314$ J/mol.K); T is the temperature (K); n is the number of electrons transferred in the half reaction; F is the Faraday's constant ($F = 96,484.6$ C/mol); Q is the reaction quotient, which is the ratio of activities of the species ($Q = \frac{[R]}{[O]}$, where $[R]$ is the surface concentration of the reductant species and $[O]$ is the surface concentration of the oxidant species).

Cyclic voltammetry (CV) and linear sweep voltammetry (LSV) have received great interest as they can be used for explaining the processes and redox mechanisms occurring on the electrode. Differential pulse voltammetry (DPV) and square wave voltammetry (SWV) are particularly useful in the determination of trace amounts of electroactive compounds in pharmaceuticals and biological fluids. Stripping voltammetry has also been widely utilized due to its ability to detect pre-concentrate analytes ultra-sensitively [48].

3.2.1 Cyclic Voltammetry

CV is a direct current (DC) potentiostatic technique where potential is swept between two limits at a constant rate and the current at the working electrode is monitored during the potential scan. The limits are chosen such that they represent the potential at which the species is expected to oxidize/reduce. The relationship between potential and time is described by a triangular waveform (Fig. 3.8).

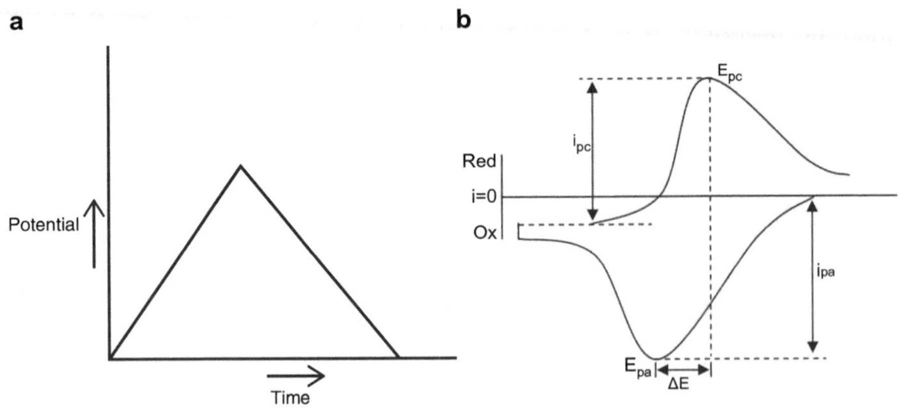

Fig. 3.8 (**a**) Typical excitation waveform for analog CV, (**b**) Theoretical CV for a reversible reaction of 1 mM $K_3[Fe(CN)_6]$ in 1 M NaCl

The voltammogram is a display of current versus potential (Fig. 3.8). After the initial non-faradaic region, the neutral species start to oxidize as the potential becomes more positive. As oxidization occurs on the electrode, the mass transfer rate of neutral species towards the electrode increases, and so does the current. This current reaches a maximum peak, after which the surface concentration of the neutral species starts to decline due to their depletion from the solution, resulting in a reduced oxidation current [Eq. (3.5)]. The process is then repeated in the same manner for the reverse scan to obtain the reduction behavior [Eq. (3.6)].

$$[Fe^{II}(CN)_6]^{3-} + e^- \rightarrow [Fe^{III}(CN)_6]^{4-} \tag{3.5}$$

$$[Fe^{III}(CN)_6]^{4-} \rightarrow [Fe^{II}(CN)_6]^{3-} + e^- \tag{3.6}$$

Activation potentials are the range of potentials at which the reactants oxidize/reduce without influences associated with diffusion/transport. Within this range, there is a steep increase in current due to the conversion of the reactants at the electrode surface. This current, also known as the "faradaic current," is the result of direct electron transfer to or from the analyte. Eventually the reaction is limited by the diffusion of reactants towards the surface, or that of products away from the surface, when the current reaches a maximum and then tails off.

The peak current (i_p), in ampere (A), is measured relative to the baseline extrapolated from the baseline signal, so as not to include charging current in the analysis.

The formal reduction potential E_0 for an electrochemically reversible couple [Eq. (3.7)] can be found at the value that corresponds to 85% of the peak potential and the thickness of the diffusion layer[3] can be estimated from $(Dt)^{1/2}$, where D is the diffusion coefficient and t is time in seconds [49]. Potential is measured in volts (V),

$$E_0 = E_{\text{pa}} + E_{\text{pc}} \tag{3.7}$$

where E_{pa} is the anodic and E_{pc} is the cathodic peak potential. For a reversible redox couple, the number of electrons transferred in the electrode reaction can be determined by estimating the separation between the peak potentials [Eq. (3.8)]:

$$\Delta E = E_{\text{pa}} - E_{\text{pc}} \cong \frac{0.059}{n} \tag{3.8}$$

[3]The diffuse-layer model depicts the distribution of net charge near the electrode surface which diminishes as the interfacial region stretches towards the bulk. The random motion of ions results in a diffuse layer of charge in which the concentration of counter ions is greatest next to the electrode surface and decreases progressively until a homogenous distribution appears in bulk electrolyte. Specific adsorption of dissolved ions at the electrode is weak and therefore the adsorbed inner layer of negative charge only partially shields the positive charge at the electrode. As a result, a layer of solvated anions is present adjacent to the diffuse layer.

CV has the capability of rapidly observing the redox behavior over a wide potential range. The potential cycle may be repeated to obtain additional information about the stability of the system. The step or scan rate can also be used to extract further information pertaining to mass transfer profile near the electrode, as the thickness of the diffusion layer can be adjusted by the scan rate.

It should be noted that the choice of the electrolyte affects the intrinsic behavior of the redox couple. The activity of the reactants also changes depending on the electrolyte concentration.

In a system where the electron transfer occurs much faster than the diffusion of an electroactive species to the electrode, the peak current, i_p (A) for the forward sweep of the first cycle, can be calculated from the Randles–Sevcik relationship [Eq. (3.9)]. Therefore, CV is useful for quantitative experiments because the current signal is directly proportional to the concentration of the analyte being measured,

$$i_p = k.n^{3/2}.A.D^{1/2}.C.v^{1/2} \quad (3.9)$$

where k is a constant; n is the number of moles of electrons transferred; A (cm^2) is the area of the electrode; D (cm^2/s) is the diffusion coefficient; C (mol/cm^3) is the solution concentration, and v (V/s) is the scan rate.

This relationship can be used to determine whether a reaction is diffusion limited or not, by simply plotting the peak current versus the square root of the scan rate. If a linear plot is obtained, then the reaction is mass transfer limited. The slope of the resulting line is proportional to the diffusion coefficient, as shown in the Randles–Sevcik equation [Eq. (3.9)]. Moreover, the peak potentials at varying scan rates can be compared to determine the reversibility of a reaction. In a reversible reaction where the electron transfer rate is fast, all peaks should appear at the same potential at different scan rates (Fig. 3.9).

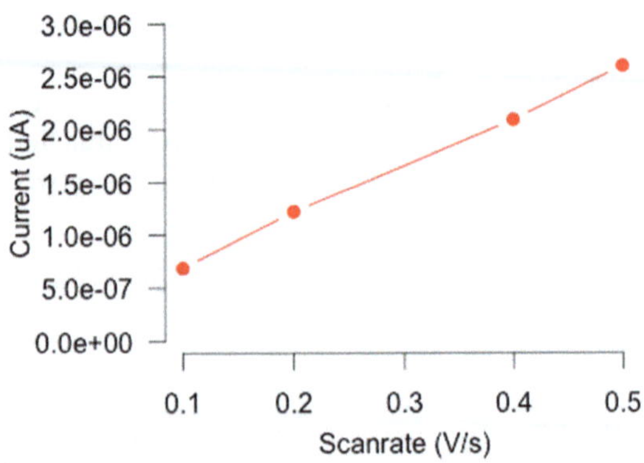

Fig. 3.9 Plot of peak height for $K_3[Fe(CN)_6]$ reduction at different scan rates

3.2.2 *Differential Pulse Voltammetry*

In order to increase speed and sensitivity, different forms of potential modulation (other than just a simple staircase ramp) have been applied over the years. DPV is an electrochemical technique where the current is measured as a function of time and the potential between the auxiliary and reference electrodes. DPV can be considered as a derivative of linear sweep voltammetry or staircase voltammetry, with a series of regular voltage pulses superimposed on the potential linear sweep or stair-steps. The current is measured immediately before each potential change, and the current difference is plotted as a function of potential.

3.2.3 *Chronoamperometry*

Chronoamperometry (CA) is a useful tool for determining diffusion coefficients and for investigating kinetics and mechanisms. Unlike the CV technique, CA can yield this information in a single experiment (Fig. 3.10).

When a potential step large enough to cause an electrochemical reaction is applied to an electrode, the current changes with time (t). The study of this current response as a function of time is called CA and is studied using the Cottrell equation [Eq. (3.10)]:

$$i = \frac{n.F.A.D^{1/2}.C^b}{(\pi.t)^{1/2}} \tag{3.10}$$

where n represents number of electrons; F, the Faraday's constant; D, the diffusion coefficient; C^b, the concentration of the bulk; and A, the surface area of the electrode.

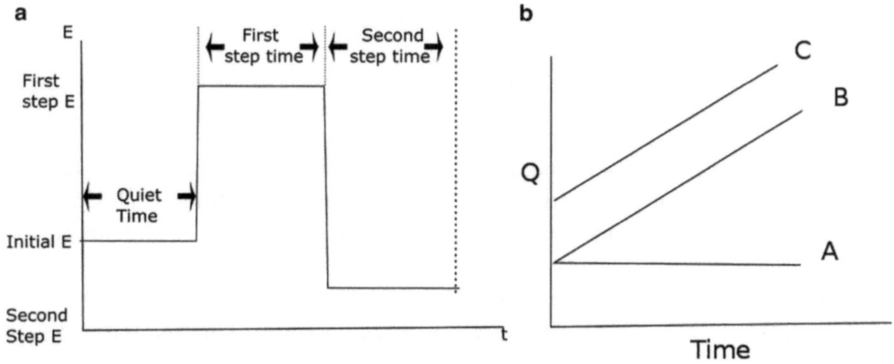

Fig. 3.10 (**a**) Potential wave form for chronoamperometry and chronocoulometry. (**b**) Theoretical Anson plots for a blank (*A*), a redox system (*B*), and a system showing adsorption (*C*)

Simply stating, the Cottrell equation shows that the current is proportional to $\frac{1}{t^{1/2}}$. It also helps calculating the diffusion coefficient by simply rearranging the equation, providing the number of electrons obtained from the CV experiment, and providing values for the other variables. Another useful parameter to calculate is the "true" area of the electrode, where the geometric area of an electrode is not the same as the electrochemically active area due to surface roughness.

3.2.4 Chronocoulometry

Chronocoulometry (CC) is an outgrowth of the chronoamperometric technique with two main advantages:

- It helps determine the kinetic rate constant more accurately,
- It helps detect adsorption of a species on an electrode surface more easily.

In a CC experiment, the charge used in an oxidation or reduction process is measured and then plotted with respect to time. The excitation waveform is the same as that used for CA, but the resulting curve shows an increase of charge with time, rather than the decrease seen in CA.

Like the CA response curve, the CC curve shows that the diffusion of the reacting species to the electrode takes some time in most reactions. This implies that at the start of the experiment (time zero) the only charge measured is due to the charging of the double layer of the electrode. In cases where a species is adsorbed on the surface of the electrode, there would be a rise in current at the initial point to larger values than would be expected for a non-adsorbed species. The technique, therefore, could also be used to calculate the thickness of the adsorption layer.

Since CC is based on the same assumptions as CA, the charge at any time can be calculated by simply integrating the Cottrell equation and adding corrections for the charge due to the double layer and other interfacial interactions (for example, adsorption). The resulting equation is [Eq. (3.11)]:

$$Q = \frac{2n.F.A.C_{\mathrm{b}}.D^{1/2}.t^{1/2}}{\pi^{1/2}} + Q_{d1} + Q_1 \tag{3.11}$$

where Q is charge; n represents number of electrons in the reaction; F, the Faraday's constant; A, the electrode surface area; C_b, the bulk concentration; D, diffusion coefficient; Q_{d1}, charge due to double layer; and Q_1, charge due to interfacial interaction.

If Q is plotted versus $t^{1/2}$, the result is known as the Anson plot. The slope of the plot would be $\frac{2n.F.A.C_{\mathrm{b}}.D^{1/2}}{\pi^{1/2}}$, if no adsorption is assumed to occur and the intercept would be the charge of the double layer. If adsorption does occur, the line in this plot would be offset by the charge due to this interfacial phenomenon. It is easy to differentiate between the charge due to the double layer and that due to adsorption (Fig. 3.10).

3.2.5 *Electrochemical Impedance Spectroscopy*

Originally developed by Sluyters et al., EIS techniques allow for the characterization of complex processes and interfaces by observing the system's impedance response to a small periodic alternating perturbation at steady state or a voltage perturbation [50]. In other words, EIS is highly sensitive for measuring changes in the electrode surface, and allows complex bio-recognition events to be probed with a simple, label-free, and mediator-free strategy.

During an impedance measurement, a small amplitude AC signal is imposed to excite the surface and the current with respect to the polarizing potential is monitored with the capacitive [imaginary (Z")] and faradaic [real (Z')] components of impedance being quantified and displayed as Nyquist plots to determine the resistive, capacitive, and inductive behavior of the surface at that particular frequency [51] (Fig. 3.11).

From these Nyquist curves, several key parameters such as solution resistance, double layer capacitance (or constant phase elements formation), and electron transfer resistance are deduced and used to measure solution stability, surface properties, or diffusion issues.

Impedance changes between electrode surfaces and a surrounding solution upon a binding event can be transduced into an electrical signal, using a frequency response analyzer; the origin of these changes, however, is hard to identify. One theory hypothesizes that binding of larger antigens forms a resistive barrier, causing

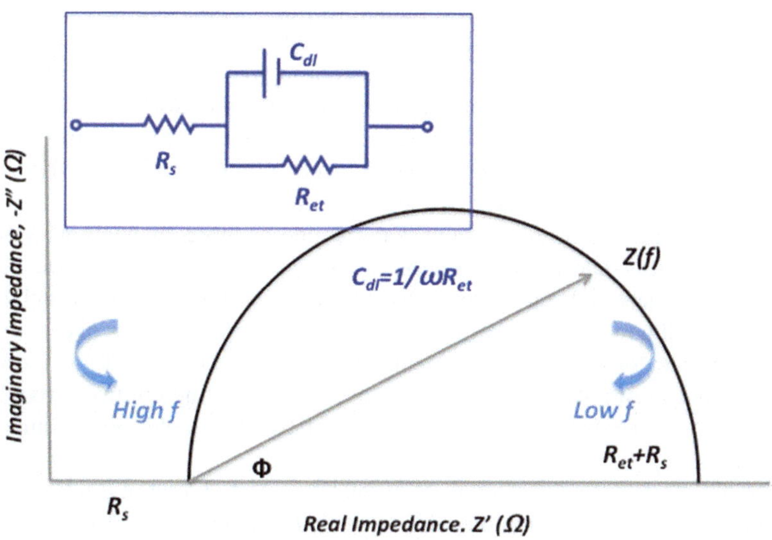

Fig. 3.11 Impedance spectra, Nyquist curve, a plot of the real impedance versus imaginary impedance. The graph is semicircular with higher frequencies associated near the origin and lower ones away from the origin

the impedance to increase, whilst binding of smaller antigens can facilitate a charge transfer and lower impedance values [52]. Another theory suggests that when the Ab/Ag complex is formed, the binding events between the hyper-variable loop regions cause conformational and thus potential changes into the system.

During EIS, a signal in the form below is applied [Eq. (3.12)]:

$$V_{(t)} = V_m . \sin(\omega . t) \tag{3.12}$$

where $V_{(t)}$ is the overall signal with an amplitude (V_m) and an angular frequency (ω). The resulting current response is measured through Eq. (3.13).

$$i_{(t)} = I_m . \sin(\omega . t + \theta) \tag{3.13}$$

The impedance at a particular frequency is then calculated simply by the following relationship [Eq. (3.14)]:

$$Z_\omega = \frac{V(t)}{i(t)} \tag{3.14}$$

3.3 Basic Concepts in Microfluidic Devices

In the past 10 years, microfluidic technology has evolved from an intriguing concept to a vigorous industry that produces numerous publications and patents each year but still is considered in its infancy. Despite the impressive laboratory results, microfluidic immunoassays have yet to make a major impact on the commercial market. A reason for this is that the necessary technologies such as fabricating structures, modifying surfaces, integrating detection, and the chemistry of the immunoassay need to be combined and optimized in order to compete with existing methods that have been perfected over decades [53].

A microfluidic device is a system that has one or more channels with at least one dimension less than 1 mm. Common fluids used in microfluidic devices include blood samples, bacterial cell suspensions, protein or antibody solutions, and various buffers. Microfluidic technology in general seeks to improve analytical performance by reducing the consumption of reagents, decreasing the analysis time, increasing reliability and sensitivity through automation, and integrating multiple processes in a single portable device. These features are particularly suitable for immunoassay applications.

Microfluidic systems for biosensing normally consist of a set of fluidic operation units that allow different biomolecules to be detected and assayed in an easy and flexible manner. Overall, the chip-based platform which has good integration with micro/nano-fluidic components is capable of sampling, filtration, preconcentration, separation, restacking, and detection of biomolecules.

Based on their flow type, microfluidic systems can be categorized into two main types, continuous and discrete.

- Continuous microfluidic systems: Continuous-flow microfluidic operation is a promising approach for well-defined and simple biochemical applications as well as chemical separation, as it is easy to implement and less sensitive to protein fouling problems. They, however, are less suitable for tasks requiring a high degree of flexibility or complicated fluid manipulations. These closed-channel systems are inherently difficult to integrate and scale because the parameters that govern the flow field vary along the flow path, making the fluid flow dependent on the properties of the entire system.
- Discrete microfluidic systems: Droplet-based microfluidic systems are currently an emerging area of microfluidic research. Commonly it benefits from the injection of multiple laminar streams of aqueous reagents into an immiscible carrier fluid, which induces flow instability needed for forming the droplets [54].

  There are several distinctive advantages for droplet-based microfluidic systems.

  - These systems are considered as a new high throughput technology that enables the generation of microdroplets in excess of several thousand per seconds.
  - They allow parallel and serial in-vitro compartmentalization.
  - In this technology, the reagents are confined inside the droplets in water-in-oil (w/o) emulsions and thus reagent transport occurs with no dispersion. This unique feature enables chemical reaction indexing, and thereby facilitates certain chemical reactions in a highly organized manner.
  - Fast mixing can occur within minute volumes of microdroplets (nanoliter to femtoliter range) due to the short diffusion distance and chaotic mixing within droplets with the use of twisting channel geometries by stretching, folding, and reorienting fluid.
  - The variation of the channel dimensions can regulate the droplet volumes. With a control of flow rate, the reagent concentrations can be modified accordingly.

While heterogeneous immunoassays benefit from the high surface area/volume ratio in the microchannel, homogeneous assays typically take advantage of the multiplexing, and very fast electrophoretic separations made possible by the microchip format.

Microfluidic devices can be used to obtain a variety of interesting measurements including molecular diffusion coefficients, fluid viscosity, pH, chemical binding coefficients, and enzyme reaction kinetics. Other applications include capillary electrophoresis, isoelectric focusing, immunoassays, flow cytometry, sample injection of proteins for analysis via mass spectrometry, polymerase chain reaction (PCR) amplification, DNA analysis, cell manipulation, cell separation, cell patterning, and chemical gradient formation. Many of these applications have utility for clinical diagnostics.

In pursuit of these goals, special interest should be focused on the following areas [55]. It should be noted that considering the final goal of this work, this section is focused on microfluidic immunoassays.

- Materials. In addition to practical considerations for fabrication, the nature of the exposed surface in the channel is also of great importance because of the large surface area to volume ratios associated with these dimensions. Given the cost advantage of plastics, which is especially important in disposable devices, the search for improved plastic materials for the microfluidic chips continues.
- Surface modification. Non-specific adsorption is one of the main challenges when working in the microscale. Modification of the surface of the channels is often necessary to prevent adsorption of the sample and key reagents, which degrades assay performance. Non-specific adsorption of biological components, especially on plastic, can also create a background signal that determines the detection limit for an assay. Surface modification is also used to improve electroosmotic flow, which requires a charged surface, in channels made of plastic for capillary electrophoresis(CE)-based immunoassays.[4]
- Introducing sample/reagent and moving fluid within the microfluidic. A strategy must be devised to interface the macroscale of the traditional laboratory to the microscale of the device.

  The capability for simultaneous multianalyte analysis could occur, probably through more complicated multipath microfluidic systems, by integration with microarrays, or by coded microbeads that carry a number of different capture antibodies.
- Microfluidic design. Design is critical to the device performance characteristics, particularly in accommodating or exploiting the properties of the fluid on the microscale.
- Detection. Detection on small volumes will improve. This is while different assay techniques use various signal transduction pathways, adapted to the very small volumes, to recognize an antibody/antigen binding event.
- Integration and miniaturization. Lack of full integration of on-chip components such as detection equipment has limited the implementation of microfluidics technology for applications outside the laboratories or research facilities. Common detection strategies involve microscopes, potentiostats, photomultiplier

[4]Microchip CE-based immunoassays: In miniaturized devices using electrophoresis to perform immunoassays, the assay relies on a difference in the electrophoretic mobilities of the immunocomplex of antigen and labeled antibody from either of the two components [56]. Thus, microchip CE assays use a separation step, electrokinetically separating the immunocomplex from the free antibody. Chip-based microfluidic immunoassays have several advantages over conventional CE assays. The channel is very small relative to the bulk of the chip, so heat is dissipated much faster than capillary-based systems. This allows higher field strengths to be used, resulting in faster separations and higher resolution. The faster separations, combined with greater capability for integrated sample preparation and the potential for multiplexing, make chip-based CE a valuable tool for high throughput applications such as combinatorial chemistry that only supply small sample volumes.

tubes, lasers, and other hardware that is bulky, expensive, and requires a trained technician to maintain. Not many completely on-board systems have been reported [57].
- Reagent stability. When portable devices are designed to be used in the field, issues of reagent stability and storage become more important, especially because the key bio-reagents are fragile.

Some of these specific areas are discussed in detail below.

3.3.1 Materials for the Microfluidic Devices

The materials used to construct microfluidic immunoassay devices vary depending on their applications, but the vast majority of such devices are constructed of glass, silicon, or polymers [58]. Each of these materials has its own advantages and limitations.

The glass materials are preferred partly for their biocompatibility and their ideal surface characteristics where high temperature or strong solvents are used [59]. However, lack of optical transparency at wavelengths of interest (for silicon), micromachining difficulties, and comparably high expenses (for both silicon and glass materials) have hampered their wider applications in microfluidics. Tremendous effort has been made to find alternative materials that are more cost-effective and easier for micromachining.

With the development of related fabrication techniques in recent years, the polymer/plastic-based microfluidic systems have garnered more interest than its conventional competitors [60]. In spite of comparatively weak bonding and structure deformation during device packaging processes, polymer materials still seem attractive as they are more cost-effective, easier to be fabricated in/on, avoidance of high-temperature annealing and stringent cleaning, more system integration friendly (e.g., interconnections). Moreover, there exists a wider range of materials to be chosen for characteristics that are required for each specific application, such as good optical transparency, biocompatibility, and chemical or mechanical properties. Another important reason for the interest from both academia and industry in polymer microfluidic devices is the possibility of fabricating disposable microfluidic chips for biomedical and clinical applications.

One particular polymer that is extensively used in this regard is polydimethylsiloxane (PDMS). PDMS is a transparent, elastomeric polymer that can be fabricated rapidly with features as small as 10 nm in dimension [61]. The elastomeric nature of PDMS helps it seal well; this is while often the adhesion due to conformal surface contact with a smooth, flat surface is enough to seal microchannels for low pressure applications. Alternatively, PDMS can be bonded irreversibly to itself, glass, or silicon simply by treating both bonding surfaces with plasma treatment.

Another significant advantage of PDMS over glass, silicon, and many other polymers is that the PDMS fabrication process can be done under normal laboratory conditions, overcoming the need for a clean room that can be expensive to maintain or difficult to access [62]. The principal limitation of PDMS for immunoassay applications is its hydrophobic surface properties. In other words, proteins and other molecules tend to adsorb to an untreated PDMS surface. Moreover, the surface is poorly wetted and thus prone to forming bubbles in aqueous systems. These limitations can be reduced or eliminated through chemical or plasma treatment of the surface [63].

Although PDMS now has an established role in prototyping new designs for microfluidic devices, its usefulness for commercial applications is limited as the fabrication process does not lend itself to the mass production accessible to injection molded or embossed hard plastics [64]. Also, PDMS is not as durable as the more rigid plastics and polymers such as polystyrene, polycarbonate, and polymethyl methacrylate (PMMA) that are often used in commercial devices.

Thermoplastic polymers, including PMMA and cyclic olefin copolymer (COC), possess low surface free energy and thus are be hydrophobic or weakly hydrophilic, limiting the strength of bonds which may be formed between mating substrates [65].

Thermoplastic polymers have several advantages including the low material cost, availability of range of material properties tailored for specific need, and amenability for high throughput fabrication. They are receiving more attention for their structural strength, optical clarity, and biocompatibility.

COC is a comparatively new thermoplastic material, and has been gaining more popularity in LOC (lab-on-chip) fabrication due to its better chemical properties [66]. COC is a copolymer consisting of ethylene and norbornene that is synthesized using a metallocene catalyst [60]. Thanks to its characteristic molecular structure and tailored catalyst technology, COC offers a wide range of grade variations, in terms of flow properties and heat resistance, such as:

- Topas-8007 (Tg = 78 °C)
- Topas-5013 (Tg = 130 °C)
- Topas-6015 (Tg = 160 °C)
- Topas-6017 (Tg = 178 °C).

As a result, depending on the specific application, an appropriate substrate material can be selected for the microdevice. Moreover, due to the olefinic properties, COC has good resistance towards hydrolysis, acids, and alkalis, and most organic polar solvents such as acetone, methanol, and isopropyl alcohol (http://www.topas.com/sites/default/files/files/optical_e.pdf). It can be soluble in organic solvents such as toluene and naphtha. Moreover, COC has high purity, good biocompatibility, and excellent optical properties.

It has higher transparency in visible and near ultraviolet regions compared to PMMA. The refractive index of COC is higher than that of PMMA, making COC ideally suited for optical detection methods. COC is easier to emboss, which is especially important when using soft PDMS tools because the density of COC is lower than that of PMMA. The water absorption of COC is 0.01%, which is ten times less than that of PMMA. Thus, the relative humidity changes in the environment will not significantly affect the devices replicated in COC.

Current methods for fabrication of microfluidic devices include prototyping techniques (hot embossing, injection molding, and soft lithography) and direct fabrication techniques such as laser photoablation or laser micromachining, photolithography/optical lithography, and X-ray lithography [67, 68]. In fact, during the past decade photolithography techniques have coupled to various plastic/polymer-based techniques to better suit LOC applications and achieve smaller feature patterning abilities. To date, most of the current soft lithography processes rely on modern photolithography techniques for master template/mask fabrication. Consequently, the low resolution ability of soft lithography can be improved with high quality masks by modern photolithography. Sub-100-nm fabrication resolution can also be achieved by composite layers of stamps [69].

Bonding can also be enhanced through the generation of electrostatic interactions, and surfaces possessing high specific energy in the form of polar functional groups, which can produce hydrogen or covalent bonds across the interface, are capable of providing bond strengths exceeding that of the bulk polymer [70].

3.3.2 *Surface Modification*

In the microscale environment, the high surface area/volume ratio magnifies the effects of non-specific binding. This is especially important in immunoassays, where key reagents such as antibodies and enzyme labels can adsorb to hydrophobic surfaces, seriously degrading the assay performance. To address this, different strategies have been applied to block or modify the surface of the microchannel, the most common of which has been the use of a blocking agent such as bovine serum albumin (BSA) to pre-coat the surface. While sufficient for many applications, the surface modification is not permanent, and the resulting surface is somewhat heterogeneous [71].

As mentioned earlier, heterogeneous assays are distinguished by attaching the primary antibody to a solid support, and a separation step that isolates the desired analyte from any other potential source of signal. The majority of heterogeneous immunoassay applications use antibodies immobilized on the microchannel surface or particles such as beads as the solid phase. Devices that use the microchannel surface for antibody immobilization have the advantage of a higher surface area to volume ratio than a standard well format.

This is while the antibodies that are directly adsorbed on a hydrophobic surface through hydrophobic effects may partially or completely denaturate due to the conformation changes associated with the adsorption process. A further limitation of immobilization on a hydrophobic surface is that blocking steps are usually necessary to limit the non-specific binding of proteins and other small molecules that could reduce the effectiveness of the assay. Different strategies to immobilize antibodies on the surface are discussed earlier in Chap. 2.

3.3.3 Introducing Sample/Reagent and Moving Fluid Within the Microfluidic

A strategy must be devised to interface the macroscale of the traditional laboratory to the microscale of the device. Reagents are typically introduced through pressure, electrokinetic, or capillary driven flow.

3.3.3.1 Pressure Driven Flow

In pressure driven flow, the fluid is pumped through the device via positive displacement pumps, such as syringe pumps. One of the basic laws of fluid mechanics for pressure driven laminar flow, known as no-slip boundary condition, states that the fluid velocity at the walls must be zero, resulting in a parabolic velocity profile within the channel. The parabolic velocity profile has significant implications for the distribution of molecules transported within a channel. Pressure driven flow can be a relative inexpensive and quite reproducible approach to pumping fluids through microdevices.

3.3.3.2 Electrokinetic Flow

Electro-osmotic pumping is another common technique for pumping fluids. In devices that use electrokinetic flow, reagents are typically pipetted or otherwise transferred into a reservoir on the device, at which point flow is controlled by applying electric potentials. Similarly, samples introduced by capillary flow are typically loaded into a reservoir connected to the microchannel.

If the walls of a microchannel have an electric charge, an electric double layer of counter ions will form at the walls. When an electric field is applied across the channel, the ions in the double layer move towards the electrode of opposite polarity, creating motion of the fluid near the walls and transfers via viscous forces into convective motion of the bulk fluid. If the channel is open at the electrodes, as is most often the case, the velocity profile is uniform across the entire width of the channel. However, if the electric field is applied across a closed channel (or a backpressure exists that just counters that produced by the pump), a recirculation pattern forms in which fluid along the center of the channel moves in a direction opposite to that walls. In closed channels, the velocity along the centerline of the channel is 50% of the velocity at the walls.

One of the advantages of electrokinetic flow is that the blunt velocity profile avoids many of the diffusion nonuniformities that occur with pressure-driven flow. However, sample dispersion in the form of band broadening is still a concern for electro-osmotic pumping. While it can easily couple with other electronic applications on-chip, electrokinetic flow often requires very high voltages, making it a difficult technology to miniaturize without off-chip power supplies. Another significant disadvantage of electrokinetic flow is the variability in surface properties.

3.3.3.3 Capillary Flow

Capillary forces result from the interaction of liquid, gas, and solid surfaces, at the interface between them. In the liquid phase, molecules are held together by cohesive forces. In the bulk of the liquid, the cohesive forces between one molecule and the surrounding molecules are balanced. However, for the same molecule at the edge of the liquid, the cohesive forces with other liquid molecules are larger than the interaction with air molecules. As a result, the liquid molecules at the interface are pulled together towards the liquid. The overall effect of these forces is to minimize the free surface of the liquid that is exposed to air. The proportionality between the decrease in energy of the surface that results from decreasing the surface is described by the surface tension.

The flow of a fluid through a microfluidic channel can be characterized by the Reynolds number, defined as Eq. (3.15), where v is the maximum velocity of the object relative to the fluid (m/s); L is a characteristic linear dimension (traveled length of the fluid) (m); μ is the dynamic viscosity of the fluid [Pa.s or N.s/m^2 or kg/(m.s)]; ν is the kinematic viscosity ($\nu = \frac{\mu}{\rho}$) (m^2/s); ρ is the density of the fluid (kg/m^3) [72].

For many microchannels, L is equal to $4A/P$ where A is the cross-sectional area of the channel and P is the wetted perimeter of the channel. Due to the small dimensions of microchannels, the Re is usually much less than 100, often less than 1.0. In this Reynolds number regime, flow is completely laminar and no turbulence occurs. Laminar flow provides a means by which molecules can be transported in a relatively predictable manner through microchannels. The transition to turbulent flow generally occurs in the range of Reynolds number 2000.

$$\text{Re} = \frac{\text{inertial forces}}{\text{viscous forces}} = \frac{\rho.v.L}{\mu} = \frac{v.L}{\nu} \tag{3.15}$$

3.3.4 *Detection*

Different assay techniques use different signal transduction pathways to recognize an antibody/antigen binding event. Most applications use a label to increase the sensitivity of detection. Current labels include metals, redox labels, optical labels, and enzymes. The majority of detection techniques mentioned in the biosensor section can similarly be adopted in microfluidic devices.

In recent years, microfluidic systems fabricated using micro-manufacturing techniques integrated with electrochemical sensing electrodes and microfluidic control devices have attracted considerable interest and shown their great potential for chemical analysis and biomedical applications.

Electrochemical detection strategies commonly have an innate high sensitivity and simplicity that can be effectively used in miniaturized hardware. As such, they constitute, arguably, the most practical, quantifiable, and saleable of all low cost diagnostic assessments of a protein's presence.

From among different types of electrochemical techniques mentioned in the previous section, amperometry, the most popular electrochemical detection method in microfluidic immunoassay systems, is a simple technique and thus miniaturization and integration can be easily achieved by creating a three-electrode system in the microfluidic chip [73]. Amperometric detection method has some advantages including simple, low background noise. However, it is not a universal method, as it requires the analyte to have the electrochemical activity; otherwise some additional substances with the electrochemical activity must be introduced to the immunoassay system for the indirect detection.

Potentiometric detection methods have the specificity because they use the ion-selective membrane to achieve charge separation. However, the selective membrane is hard to integrate into the microfluidic system. Conductometric detection method is a universal detection method but lacks specificity. Moreover, the conductivity between the background solution and the analyte must be significantly different; else they cannot be distinguished using this method.

A significant advantage of voltammetry for detection in a microfluidic system is the improvement in performance that accompanies the reduction of electrode size into the low micrometer range. As the electrode diameter is decreased into the low micrometer range, a shift to nonplanar diffusion (e.g., hemispherical) occurs, causing an increase in the collection efficiency of the electroactive species at the surface [74]. The practical result is an increase in the signal-to-noise ratio, which generally translates into a lower detection limit [75].

3.3.5 Design

The materials, shape, and complexity of different microfluidic systems vary depending on the intended application. Some devices consist of a single channel, while others have complex patterns with multiple integrated components. While the on-chip integration of components such as mixers, valves, and detectors has the advantage of increased portability, the complexity of the fabrication process increases dramatically, and the reliability of the device diminishes.

This is while, adding multiple channels or other more complex two-dimensional designs does not significantly affect the time or difficulty of fabrication, and so many high throughput multiplexing applications can be efficiently addressed using microfluidics [56].

References

1. R. Yalow, S. Berson, Assay of plasma insulin in human subjects by immunological methods. Nature **184**, 1648–1649 (1959)
2. E. Engvall, P. Perlmann, Enzyme-linked immunosorbent assay (ELISA) quantitative assay of immunoglobulin g. Immunochemistry **8**(9), 871–874 (1971)

3. J. Butler, Enzyme-linked immunosorbent assay. J. Am. Geriatr. Soc. Immunoassay **21**(2–3), 165–209 (2000)
4. J. Vaitukaitis, G. Braunstein, G. Ross, A radioimmunoassay which specifically measures human chorionic gonadotropin in the presence of human luteinizing hormone. Am. J. Obstet. Gynecol. **113**(6), 751–758 (1972)
5. S. Centi, S. Laschi, M. Mascini, Strategies for electrochemical detection in immunochemistry. Bioanalysis **1**(7), 1271–1291 (2009)
6. P. von Lode, Point-of-care immunotesting: approaching the analytical performance of central laboratory methods. Clin. Biochem. **38**(7), 591–606 (2005)
7. A. Sassolas, L.J. Blum, B.D. Leca-Bouvier, Immobilization strategies to develop enzymatic biosensors. Biotechnol. Adv. **30**(3), 489–511 (2012)
8. D.R. Thévenot, K. Toth, R.A. Durst, G.S. Wilson, Electrochemical biosensors: recommended definitions and classification. Biosens. Bioelectron. **16**(1–2), 121–131 (2001)
9. F. Faridbod, V.K. Gupta, H.A. Zamani, Electrochemical sensors and biosensors. Int. J. Electrochem. **2011**, Article ID 352546 (2011)
10. H. Nassef, L. Civit, A. Fragoso, C. O'Sullivan, Amperometric immunosensor for detection of celiac disease toxic gliadin based on fab fragments. Anal. Chem. **81**(13), 5299–5307 (2009)
11. A. Warsinke, Point-of-care testing of proteins. Anal. Bioanal. Chem. **393**(5), 1393–1405 (2009)
12. A. Koyun, E. Ahlatcioglu, Y.K. Ipek, Biosensors and their principles, in *A Roadmap of Biomedical Engineers and Milestones* (InTech, Rijeka, 2012)
13. G. Zheng, F. Patolsky, Y. Cui, W. Wang, C. Lieber, Multiplexed electrical detection of cancer markers with nanowire sensor arrays. Nat. Biotechnol. **23**(10), 1294–1301 (2005)
14. D.S. Hage, Immunoassays. Anal. Chem. **67**(12), 455–462 (1995). Available: http://dx.doi.org/10.1021/ac00108a030
15. J. Wang, A. Ibáñez, M.P. Chatrathi, A. Escarpa, Electrochemical enzyme immunoassays on microchip platforms. Anal. Chem. **73**(21), 5323–5327 (2001)
16. D.L. Stokes, G.D. Griffin, T. Vo-Dinh, Detection of E. coli using a microfluidics-based antibody biochip detection system. Fresenius J. Anal. Chem. **369**(3–4), 295–301 (2001)
17. A. Hatch, A.E. Kamholz, K.R. Hawkins, M.S. Munson, E.A. Schilling, B.H. Weigl, P. Yager, A rapid diffusion immunoassay in a t-sensor. Nat. Biotechnol. **19**(5), 461–465 (2001)
18. S. Iijima, Helical microtubules of graphitic carbon. Nature **354**(6348), 56–58 (1991)
19. H. Ke-Jing, N. De-Jun, X. Wan-Zhen, W. Wei, A disposable electrochemical immunosensor for carcinoembryonic antigen based on nano-au/multi-walled carbon nanotubes–chitosans nanocomposite film modified glassy carbon electrode. Anal. Chem. Acta **659**(1–2), 102–108 (2010)
20. M. Daniel, D. Astruc, Gold nanoparticles: assembly, supramolecular chemistry, quantum-size-related properties, and applications toward biology, catalysis, and nanotechnology. Chem. Rev. **104**, 293–346 (2004)
21. T. Sreeprasad, T. Pradeep, Noble metal nanoparticles, in *Springer Handbook of Nanomaterials* (Springer, Berlin/Heidelberg, 2013)
22. L. Hirsch, J. Jackson, A. Lee, N. Halas, J. West, A whole blood immunoassay using gold nanoshells. Anal. Chem. **75**(10), 2377–2381 (2003)
23. J. Svarc-Gajic, Z. Stojanovic, Z. Suturovic, N. Marjanovic, S. Kravic, Direct mercury determination in natural waters by chronopotentiometric stripping analysis in macroelectrode process vessel. Desalination **249**, 253–259 (2009)
24. T. Hezarda, K. Fajerwerge, D. Evrardc, V. Collièree, P. Behraa, P. Grosc, Influence of the gold nanoparticles electrodeposition method on hg(ii) trace electrochemical detection. Electrochim. Acta **73**, 15–22 (2012)
25. W. Vastarella, L.D. Seta, A. Masci, J. Maly, M. De Leo, L.M. Moretto, R. PIlloton, Biosensors based on gold nanoelectrode ensembles and screenprinted electrodes. Int. J. Environ. Anal. Chem. **87**(10–11), 701–714 (2007)
26. Y. Lia, H. Schluesenerb, S. Xua, Gold nanoparticle-based biosensors. Gold Bull. **43**(1), 29–41 (2010)

27. G. Lai, F. Yan, H. Ju, Dual signal amplification of glucose oxidase-functionalized nanocomposites as a trace label for ultrasensitive simultaneous multiplexed electrochemical detection of tumor markers. Anal. Chem. **81**(23), 9730–9736 (2009)
28. J. Pingarron, P. Yanez-Sedeno, A. Gonzalez-Cortes, Gold nanoparticle-based electrochemical biosensors. Electrochim. Acta **53**, 5848–5866 (2008)
29. T.R. Holford, F. Davis, S.P. Higson, Recent trends in antibody based sensors. Biosens. Bioelectron. **34**(1), 12–24 (2012)
30. X. Fan, I.M. White, S.I. Shopova, H. Zhu, J.D. Suter, Y. Sun, Sensitive optical biosensors for unlabeled targets: a review. Anal. Chem. Acta **620**(1–2), 8–26 (2008)
31. T. Shimomura, T. Sumiya, M. Ono, T. Ito, T.-A. Hanaoka, A novel, disposable, screen-printed amperometric biosensor for ketone 3-b-hydroxybutyrate fabricated using a 3-b-hydroxybutyrate dehydrogenase-mesoporous silica conjugate. Anal. Bioanal. Chem. **405**(1), 297–305 (2013)
32. D. Shankaran, K. Gobi, N. Miura, Recent advancements in surface plasmon resonance immunosensors for detection of small molecules of biomedical, food and environmental interest. Sens. Actuators B **12**, 158–177 (2007)
33. J. Bard, L. Faulkner, *Electrochemical Methods: Fundamentals and Applications*, 2nd edn. (Wiley, New York, 2001)
34. J. Zhu, Z. Zhu, Z. Lai, R. Wang, X. Guo, X. Wu, G. Zhang, Z. Zhang, Y. Wang, Z. Chen, Planar amperometric glucose sensor based on glucose oxidase immobilized by chitosan film on Prussian blue layer. Sensors **2**, 127–136 (2002)
35. R. Chai, R. Yuan, Y. Chai, C. Ou, S. Cao, X. Li, Amperometric immunosensors based on layer-by-layer assembly of gold nanoparticles and methylene blue on thiourea modified glassy carbon electrode for determination of human chorionic gonadotrophin. Talanta **74**(5), 1330–1336 (2008)
36. P.B. Luppa, L.J. Sokoll, D.W. Chan, Immunosensors—principles and applications to clinical chemistry. Clin. Chim. Acta **314**(1–2), 1–26 (2001)
37. J. Haun, T.-J. Yoon, H. Lee, R. Weissleder, Magnetic nanoparticle biosensors. Wiley Interdiscip. Rev. Nanomed. Nanobiotechnol. **2**(3), 291–304 (2010)
38. N.T. Beneitez, J. Missinne, E.W.A. Visser, L.J. van IJzendoorn, M.W.J. Prins, J.J.H.B. Schleipen, J.L. Vinkenborg, G. Rietjens, M.A. Verschuuren, G. Krishnamoorthy, J.G. Orsel, G. Van Steenberge, Evanescent field biosensor using polymer slab waveguide-based cartridges for the optical detection of nanoparticles. IEEE J. Sel. Top. Quantum Electron. **22**(3), 1–8 (2016)
39. D. Bruls, T. Evers, J. Kahlman, P. Van Lankvelt, M. Ovsyanko, E. Pelssers, J. Schleipen, F. De Theije, C. Verschuren, T. Van Der Wijk, J. Van Zon, W. Dittmer, A. Immink, J. Nieuwenhuis, M. Prins, Rapid integrated biosensor for multiplexed immunoassays based on actuated magnetic nanoparticles. Lab Chip **9**(24), 3504–3510 (2009)
40. J. McEwen, Potentiometric methods, in *The Dynamic Chemistry E-Textbook* (UC Davis, Davis, 2014)
41. A. Washburn, M. Luchansky, A. Bowman, R. Bailey, Quantitative, label-free detection of five protein biomarkers using multiplexed arrays of silicon photonic microring resonators. Anal. Chem. **82**(1), 69–72 (2010)
42. V. Mecea, From quartz crystal microbalance to fundamental principles of mass measurements. Anal. Lett. **38**, 753–767 (2005)
43. A.M. Coto-Garcia, E.S. Sotelo-Gonzalez, M.T. Fernandez-Arguelles, R. Pereiro, J.M. Costa-Fernandez, A. Sanz-Medel, Nanoparticles as fluorescent labels for optical imaging and sensing in genomics and proteomics. Anal. Bioanal. Chem. **399**(1), 29–42 (2011)
44. E. Gileadi, *Electrode Kinetics for Chemists, Chemical Engineers, and Materials Scientists* (Wiley, New York, 1993)
45. A.I. Mohammed, S. Gupta, S. Hussaini, A review on electrochemical biosensors: principles and applications. Adv. Biores. **3**(4), 158–163 (2012)
46. N.J. Ronkainen, H.B. Halsall, W.R. Heineman, Electrochemical biosensors. Chem. Soc. Rev. **39**(5), 1747–1763 (2010)

47. M. Thompson, L. Kateley, The Nernst equation: determination of equilibrium constants for complex ions of silver. J. Chem. Educ. **76**(1), 95–96 (1999)
48. P. Kissinger, W.R. Heineman, Cyclic voltammetry. J. Chem. Educ. **60**(9), 702–706 (1983)
49. R. Greef, R. Peat, L. Peter, D. Pletcher, J. Robinson, *Instrumental Methods in Electrochemistry* (Wiley, New York, 1985)
50. B. Chang, S. Park, Electrochemical impedance spectroscopy. Annu. Rev. Anal. Chem. **3**, 207–229 (2010)
51. S. Grant, F. Davis, K.A. Law, A.C. Barton, S.D. Collyer, S.P. Higson, T.D. Gibson, Label-free and reversible immunosensor based upon an ac impedance interrogation protocol. Anal. Chem. Acta **537**(1–2), 163–168 (2005)
52. E. Tully, S.P. Higson, R. O'Kennedy, The development of a 'labeless' immunosensor for the detection of listeria monocytogenes cell surface protein, internalin b. Biosens. Bioelectron. **23**(6), 906–912 (2008)
53. A. Bange, H.B. Halsall, W.R. Heineman, Microfluidic immunosensor systems. Biosens. Bioelectron. **20**(12), 2488–2503 (2005)
54. J. Tice, H. Song, A. Lyon, R. Ismagilov, Formation of droplets and mixing in multiphase microfluidics at low values of the Reynolds and the capillary numbers. Langmuir **19**, 9127–9133 (2003)
55. C.-Y. Liu, J. Rick, T.-C. Chou, H.-H. Lee, G.-B. Lee, Integrated microfluidic system for electrochemical sensing of urinary proteins. Biomed. Microdevices **11**(1), 201–211 (2009)
56. A. Bromberg, R. Mathies, Multichannel homogeneous immunoassay for detection of 2,4, 6-trinitrotoluene (TNT) using a microfabricated capillary array electrophoresis chip. Electrophoresis **25**(12), 1895–1900 (2004)
57. J.-W. Choi, K.W. Oh, J.H. Thomas, W.R. Heineman, H.B. Halsall, J.H. Nevin, A.J. Helmicki, H.T. Henderson, C.H. Ahn, An integrated microfluidic biochemical detection system for protein analysis with magnetic bead-based sampling capabilities. Lab Chip **2**, 27–30 (2002)
58. J. Wu, M. Gu, Microfluidic sensing: state of the art fabrication and detection techniques. J. Biomed. Opt. **16**(8), 080901 (2011)
59. P. Vulto, T. Huesgen, B. Albrecht, G.A. Urban, A full-wafer fabrication process for glass microfluidic chips with integrated electroplated electrodes by direct bonding of dry film resist. J. Micromech. Microeng. **19**, 077001 (2009)
60. R.K. Jena, C.Y. Yue, Cyclic olefin copolymer based microfluidic devices for biochip applications: ultraviolet surface grafting using 2-methacryloyloxyethyl phosphorylcholine. Biomicrofluidics **6**(1), 12822 (2012)
61. S. Sia, G. Whitesides, Microfluidic devices fabricated in poly(dimethylsiloxane) for biological studies. Electrophoresis **24**(21), 3563–3576 (2003)
62. J. McDonald, D. Duffy, J. Anderson, D. Chiu, H. Wu, O. Schueller, G.M. Whitesides, Fabrication of microfluidic systems in poly(dimethylsiloxane). Electrophoresis **21**(1), 27–40 (2000)
63. H. Makamba, J. Kim, K. Lim, K. Park, J. Hahn, Surface modification of poly(dimethylsiloxane) microchannels. Electrophoresis **24**(21), 3607–3619 (2003)
64. R. Mukhopadhyay, When PDMS isn't the best. Anal. Chem. **79**, 3248–3253 (2007)
65. C. Tsao, L. Hromada, J. Liu, P. Kumar, D.L. DeVoe, Low temperature bonding of PMMA and COC microfluidic substrates using uv/ozone surface treatment. Lab Chip **7**, 499–505 (2007)
66. T. Nielsen, D. Nilsson, F. Bundgaard, P. Shi, P. Szabo, O. Geschke, A. Kristensen, Nanoimprint lithography in the cyclic olefin copolymer, topas, a highly ultraviolet-transparent and chemically resistant thermoplast. J. Vac. Sci. Technol. B Microelectron. Nanometer Struct. **22**, 1770–1775 (2004)
67. H. Becker, U. Heim, Hot embossing as a method for the fabrication of polymer high aspect ratio structures. Sens. Actuators A **83**, 130–135 (2000)
68. U. Attia, S. Marson, J. Alcock, Micro-injection moulding of polymer microfluidic devices. Microfluid. Nanofluid. **7**, 1–28 (2009)
69. T.W. Odom, J.C. Love, D.B. Wolfe, K.E. Paul, G.M. Whitesides, Improved pattern transfer in soft lithography using composite stamps. Langmuir **18**, 5314–5320 (2002)

70. W. Niles, P. Coassin, Cyclic olefin polymers: innovative materials for high-density multiwell plates. Assay Drug Dev. Technol. **6**, 577–590 (2006)
71. V. Linder, E. Verpoorte, W. Thormann, N. de Rooij, H. Sigrist, Surface biopassivation of replicated poly(dimethylsiloxane) microfluidic channels and application to heterogeneous immunoreaction with on-chip fluorescence detection. Anal. Chem. **73**(17), 4181–4189 (2001)
72. H. Bruus, *Theoretical Microfluidics* (Oxford University Press, Oxford, 2007)
73. Y.S. Choi, Y. Ding, C.H. Ahn, H.B. Halsall, W.R. Heineman, A microchip electrochemical immunosensor fabricated using micromachining techniques, in *Proceedings of the 19th Annual International Conference of the IEEE Engineering in Medicine and Biology Society, 1997*, vol. 5 (1997), pp. 2264–2266
74. A. Michael, R. Wightman, Microelectrodes, in *Laboratory Techniques in Electroanalytical Chemistry*, 2nd edn. (Dekker, New York, 1996)
75. J. Marcus, W. Anderson, S. Quake, Parallel picoliter RT-PCR assays using microfluidics. Anal. Chem. **78**, 956–958 (2006)

Chapter 4
Conjugated AuNP-Antibody Nanoprobe Fabrication and Validation

Abstract As the first step, through the inherent interaction between gold nanoparticles (AuNPs) and antibody biomolecules, a novel gold nanoprobe was developed to be used in the non-enzymatic electrochemical immunoassay. This one-pot method not only provides a simple method for loading high-content antibody on nanoparticles but also greatly improves the repeatability and controllability of the nanoprobe preparation. Combined with a disposable electrode, it could be used for a single/multiplexed electrochemical immunosensing method. In other words, using this nanoprobe, high throughput instrumentation applying (single/multiplexed sensors) could be developed to screen a variety and vast quantities of samples for choice markers.

4.1 Introduction

An electrochemical biosensor usually consists of three parts:

- A receptor biomolecule, which incorporates a biological or biomimetic recognition element (nucleic acid, enzyme, antibody, tissue, organelles, or whole cells) and plays a key role in determining the sensitivity of the electrochemical biosensors.
- A transducer, a conductive phase, which converts biochemical signal into a measurable one based on the interaction between the analyte and the biological molecule associated with the transducer surface.
- An amplifier, a computerized apparatus/software used to boost/read-out of signal.

An efficient biosensor, therefore, relies heavily on selective/specific binding of the recognition element with the target analytes as well as the transducer for signaling the binding event in terms of response time, signal-to-noise (S/N) ratio, and limits of detection (LoD) [1]. Therefore, designing sensors with higher efficacy depends on the development of novel materials to improve both the recognition and transduction steps.

Nanomaterials, more specifically noble metal nanoparticles (NPs) have played an important role in the development of new biosensors, and/or in the enhancement of existing biosensing techniques to fulfill the demand for more specific and highly sensitive biomolecular diagnostics [2, 3]. They can measure between 1 nm and

P. Khashayar, *Point-of-Care Solution for Osteoporosis Management*,
DOI 10.1007/978-3-319-55053-4_4

100 nm in diameter, have different shapes, and can be composed of one or more inorganic/metallic compounds. The majority of them exhibit size-related properties that differ significantly from those observed in micron-sized particles or bulk materials. Recent studies have shown that the labeling of biomolecules such as antibodies with noble metal NPs results in highly sensitive and specific bioassays, while maintaining the bioactivity of the biomarker [4]. The concentration of analytes can also be detected based on the signal detection of such NPs.

Gold nanoparticles (AuNPs) are the most commonly used labels in sensor studies because of their intriguing size-dependent electrical, optical, magnetic, and chemical properties as well as their ease of synthesis, characterization, and surface functionalization [5, 6]. Being redox active, AuNPs provide the possibility for the miniaturization of sensing devices to the nanoscale, offering excellent prospects for chemical and biological sensing. They also offer a suitable platform for multifunctionalization with a wide range of organic or biological ligands for the selective binding and detection of biomarkers [7]. This means that the functional activity of these molecules remains unaffected even after their immobilization on AuNPs.

Moreover, because of their high surface area to volume ratios, AuNPs serve as an excellent scaffold to immobilize large quantities of organic or biomolecules, thus increasing the chance of their interaction with a target analyte [8]. On the other hand, AuNPs (e.g., citrate capped) are mostly negatively charged, and can therefore be tailored to electrostatically interact with certain positively charged biomolecules that in turn can have a highly selective interaction exclusively with the target analyte of interest. Furthermore, biofunctionalized AuNPs also have the advantage of multiplexing in the detection of analytes. AuNPs offer size-based signal amplification as one of the benefits and thus substantial increase in the size of these particles increases the sensitivity of the sensor.

In view of the above-mentioned facts, broad potential of AuNPs for signal amplification of antibody–antigen recognition events on simple, ultrasensitive, multiplexed immunosensors becomes apparent [9]. Thus much attention should be devoted to the preparation and application of antibody-AuNP conjugates with better operability and controllability in biomarker detection. With the developed and discussed technique herein, a simple, sensitive, and low cost solution has been modeled to be used for measuring serum levels of biomarkers later on in this work and in future in single and/or multiplexed sensor systems.

4.2 Synthesis of a Conjugated AuNP-Antibody Nanoprobe

4.2.1 Synthesis of AuNPs

Ever since Faraday's work in 1857, several methods have been applied to synthesize AuNPs with a good level of homogeneity and provide fine control over size, shape, and surface properties, in order to better take advantage of their unique

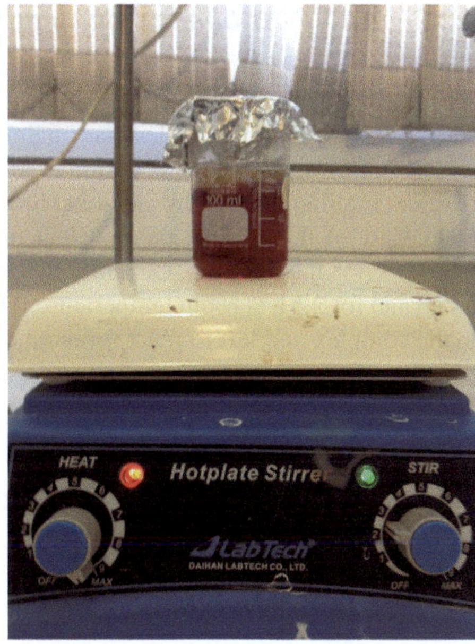

Fig. 4.1 Colloid gold preparation process

physicochemical properties for biosensing [10]. These AuNPs, also known as colloidal gold, can be easily synthesized in sizes ranging between 3 nm and 200 nm in diameter and in different shapes, being the most common the quasi-spherical shape, mainly due to their surface energy that favors the formation of spherical particles.

One of the most common approaches for the preparation of AuNPs in aqueous solution was introduced by Turkevich, and subsequently improved by Frens [11, 12]. In order to synthesize 10–100 nm AuNPs, generally, tetrachloroauric acid ($HAuCl_4$) is reduced in the presence of different concentrations of trisodium citrate as a reducing agent. Briefly, in this work, 200 ml of 0.01% $HAuCl_4$ solution was brought to boil with gentle stirring. Thereafter, 4.5 ml of 1% trisodium citrate was rapidly added and boiling was pursued for an additional 15 min. The solution was stirred continuously after removing the beaker from the hot plate. Once mixing occurred, the color of the solution changed gradually from light yellow to wine red. The solution was then cooled down to room temperature. This solution was stored at 4 °C for further use (Fig. 4.1).

The sodium citrate, in this procedure, not only behaves as a reducing agent but also provides the negative surface charge adsorbed around the AuNPs, which can repel the particles and restrain the formation of any aggregation. The concentrations of gold salt and trisodium citrate as well as temperature and mixing rate are main factors defining the size distribution of the synthesized gold nanoparticles [13]. The strength of the observed color of solution is highly dependent on the quality as well as the size of the developed colloidal gold particles [14].

4.2.2 AuNP-Antibody Conjugation

As mentioned in Chap. 2, conjugation of noble metal nanoparticles such as gold and silver with biomolecules has mainly been achieved by one of four major mechanisms: electrostatic adsorption, direct chemisorption of thiol derivatives, covalent binding through bifunctional linkers, and specific affinity interactions [15].

In view of these facts, the covalent binding through cross-linkers was used in this work. As the first step, Oc antibody (10 μg/mL) and sulfo-NHS (100 mM) were incubated for 2.5 h at room temperature. The GSH solution (10 mM) was then added to the complex and left to react for another hour. Thereafter the AuNP solution was added.

4.3 Characterization of Conjugated AuNP-Antibody Nanoprobe

4.3.1 Characterization of AuNPs

In order to obtain information about structure, morphology, and size of the synthesized AuNPs, UV-Vis spectroscopy, TEM analysis, and zeta potential measurements were applied.

The UV-Vis spectrophotometer has been the first optical technique employed for the characterization of AuNPs. In this method, the width of the absorption spectra is related to the size distribution range of the nanoparticles and with increasing particle size, the absorption peak shifts to longer wavelengths [16]. Generally, gold nanospheres display a single absorption peak in the visible range between 510 and 550 nm, mainly at 520 nm due to surface plasmon resonance properties (SPR) of colloidal AuNPs prepared by the conventional citrate reduction method [17] (Fig. 4.2). Similarly, the UV-Vis spectrum of the as-prepared AuNP showed only

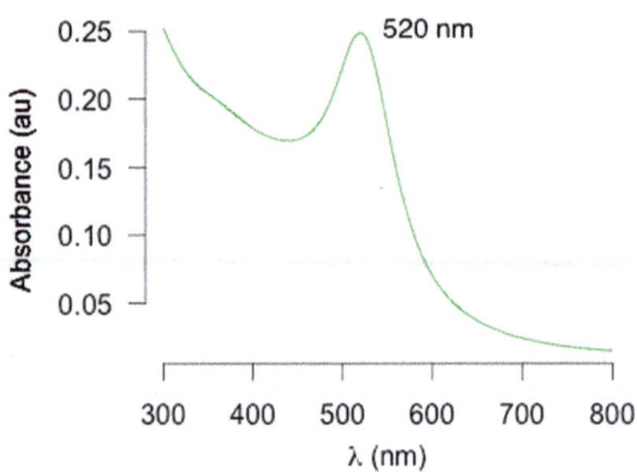

Fig. 4.2 UV-Vis spectra of colloidal AuNPs

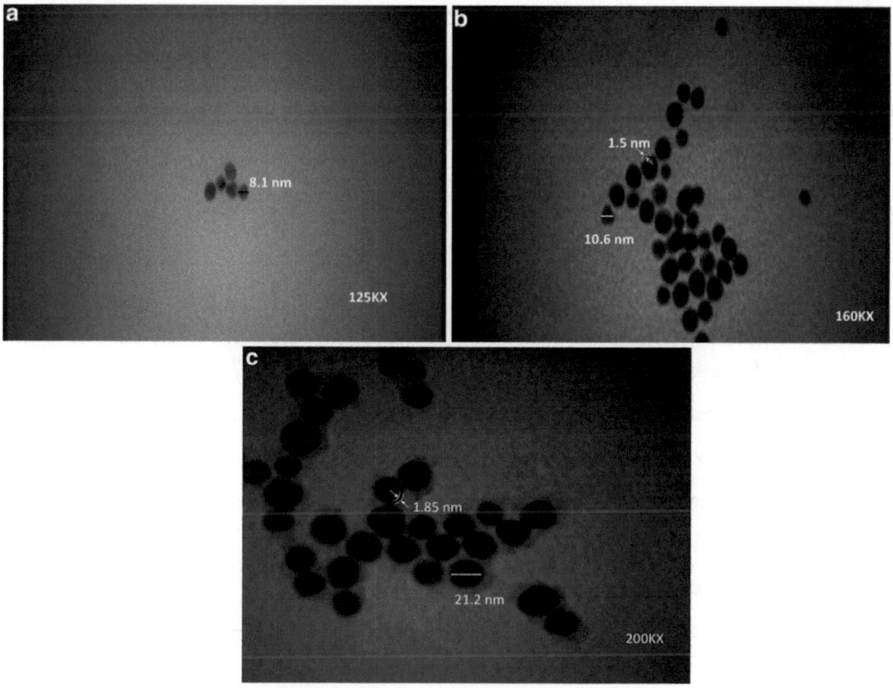

Fig. 4.3 TEM results of (**a**) monodispersed gold nanoparticles, (**b**) gold nanoparticles capped with GSH, (**c**) antibody-conjugated gold nanoparticles. (For (**a**), scale = 125 KX; for (**b**), scale = 160 KX; for (**c**), scale = 200 KX)

one band with a maximum wavelength (λ) at 520 nm. Lack of a band at around 680 nm indicates the absence of aggregated gold particles [18].

The size and size-distribution of gold nanoparticles developed using the Turkevich and Frens method is reported to be dependent on the concentrations of gold salt, trisodium citrate, temperature, and mixing rate. TEM was used to characterize the morphology and determine the size of the as-prepared AuNPs. According to the results, numerous uniformly distributed nanoparticles with an average size of 6.53 $\pm$ 1.8 nm were observed in the solution with very few particles of higher and lower size distribution. These images also show that the majority of the developed gold nanoparticles were spherical in shape (Fig. 4.3).

The hydrodynamic diameter of nanoparticles was measured using a zeta-sizer nano-ZS (Fig. 4.4). The peak mean, about 7.48 nm, shows the mean diameter of the majority of the particles according to number (%). The particles had a mean zeta potential of 42.6 mV.

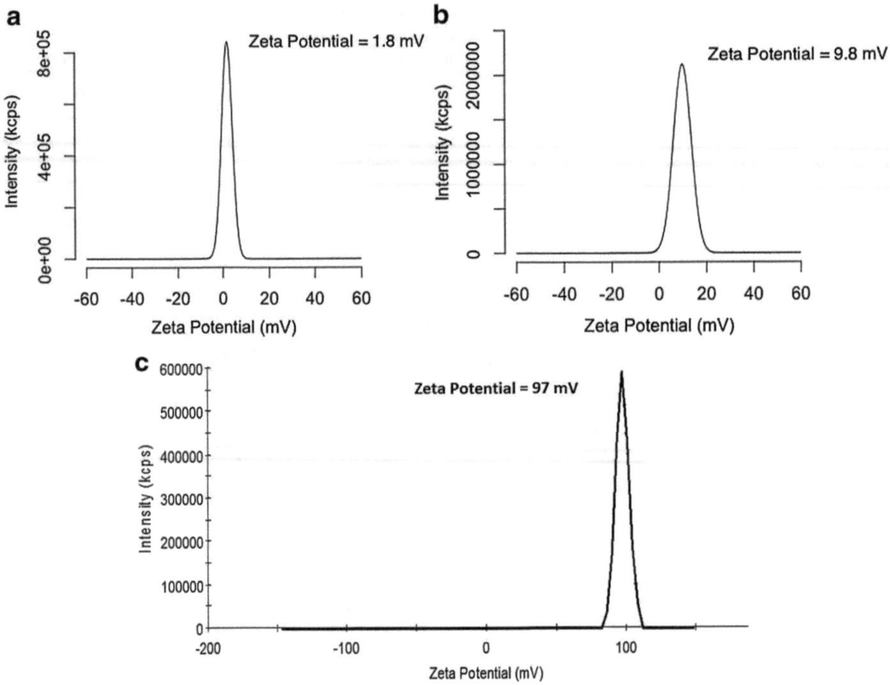

Fig. 4.4 Zeta potential results of (**a**) gold nanoparticles, (**b**) GSH-capped gold nanoparticles, (**c**) antibody-conjugated gold nanoparticles

4.3.2 Characterization of AuNP/Ab Nanocomplex

AuNP conjugation with different proteins is well studied in the literature but the mechanisms behind the process are poorly understood. Therefore special care needs to be taken throughout the process to ensure the quality of the end product. The protocol used for conjugation of AuNPs to Oc antibody in this work relies on covalent and electrostatic attractions between constituents.

In this reaction, sulfo-NHS acts as a cross-linker facilitating the bonding of an amine group of the antibody to a carboxylic group of GSH (Fig. 4.5). GSH, on the other end, provides the thiol groups needed to bind the antibody complex to AuNPs. In other words, binding of antibody complex to gold atoms happened through thiol linkages, ionic interactions between the negatively charged gold surface and the positively charged terminals on the complex, and interactions originating from absorption of hydrophobic groups on AuNPs.

Another advantage of the above-mentioned method is covalent coupling via a cross-linker being specific and controllable. The tertiary structure of the protein is maintained in this method and thus the antibody remains functional in the future steps.

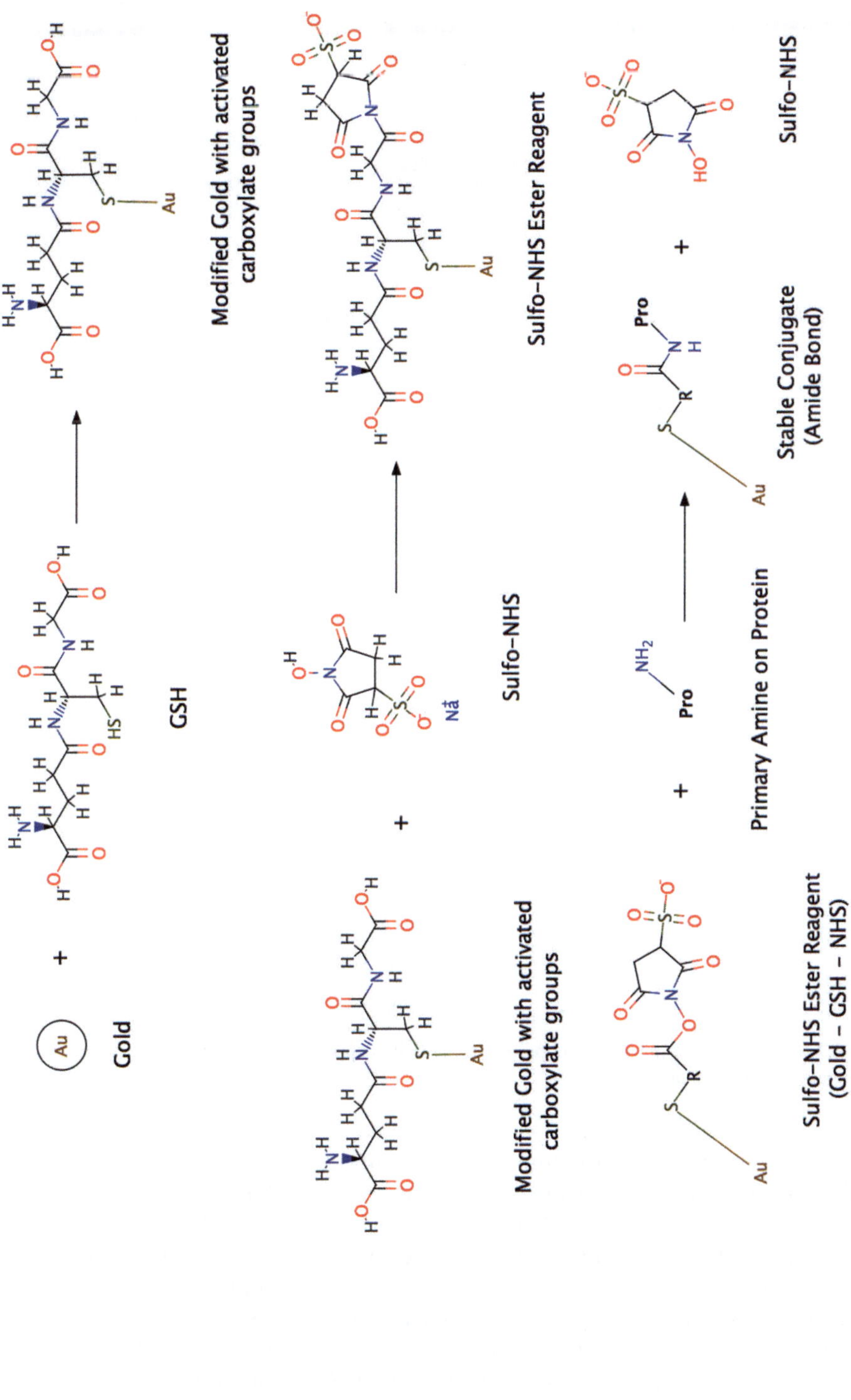

Fig. 4.5 Chemical reactions occurring in different steps

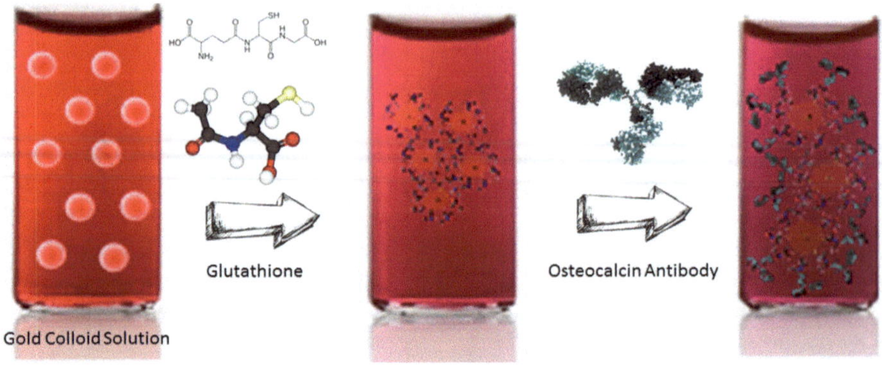

Fig. 4.6 Representative schematic showing nanoconjugate formation steps

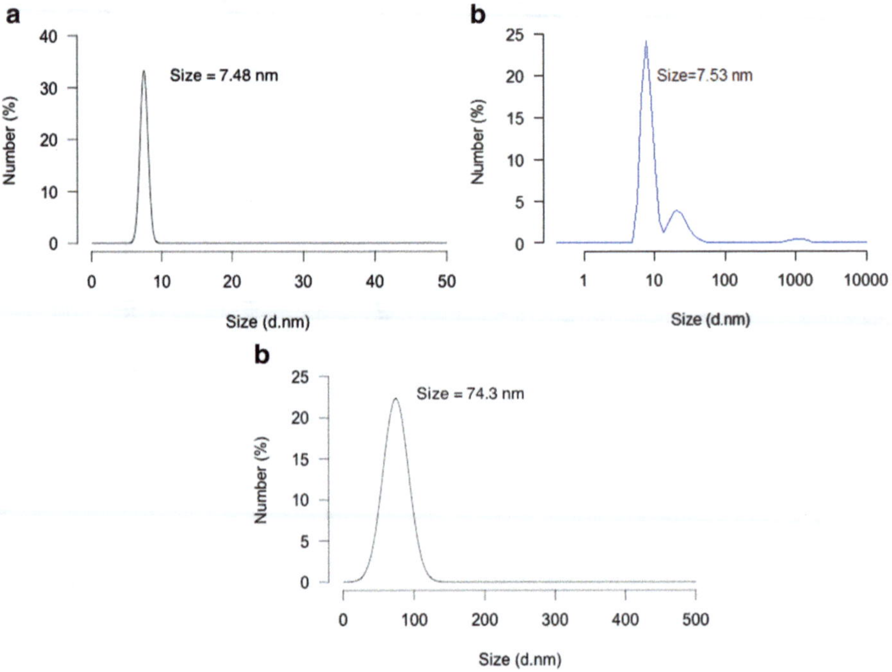

Fig. 4.7 DLS results of (**a**) gold nanoparticles, (**b**) GSH-capped gold nanoparticles, (**c**) antibody-conjugated gold nanoparticles

The conjugates developed through direct absorption of the antibody on the surface of the nano-colloid particles using non-covalent bonds, such as London, van der Waals force, and hydrophobic interactions, were not stable (Fig. 4.6).

Characterization of the resulting conjugates was done promptly by TEM measurements of freshly prepared solution (Fig. 4.3). According to these results, dispersed AuNPs became aggregated upon addition of GSH without significant alteration in particle morphology or size. This was also confirmed by DLS, which showed a slight increase in nanoparticle size after GSH addition (Fig. 4.7).

The attachment of the Ab to GSH-modified gold nanoparticles, on the other hand, resulted in a considerable increase in size and the core-shell structure, in accordance with previously reported Ab-conjugated-nanoparticles [19, 20]. According to DLS, the size of the antibody-conjugated nanoparticles was as high as 74.3 nm (Fig. 4.4). Adding GSH and Ab also considerably increased the electro kinetic potential of the nanoparticles from 1.8 mV to 9.8 mV and 97 mV, respectively.

4.4 Optimization of AuNP/Ab Nanoconjugate Preparation

The preparation of a well-dispersed nanoprobe with uniform size distribution and no aggregation is of great importance in immunoassay studies. Therefore lowest amount of antibody enough to stabilize the gold was determined by incubating 1 mL of colloid gold with antibody solutions at varying concentrations (1–50 μg/mL, 1 mL). After 5 min, 0.5 mL of 10% (w/v) NaCl was added and the color of the solution was noted. The concentration in which the color of the solution turned blue was considered the minimum required amount for stabilization.

Further tests while using the nanoprobe in an electrochemical sensor showed that lower concentrations of Ab in the nanoprobe produce small electrochemical response, suggesting that such low concentrations do not provide sufficient antibody to react completely with the antigen on the immunosensor surface (Fig. 4.8). Higher concentrations of the nanoprobe, on the other hand, improved the current while contributing to higher non-specific adsorption.

4.4.1 *Environmental Factors*

As mentioned in the literature, environmental factors such as temperature and pH play an important role in the efficiency of the nanoconjugate probe. Previous

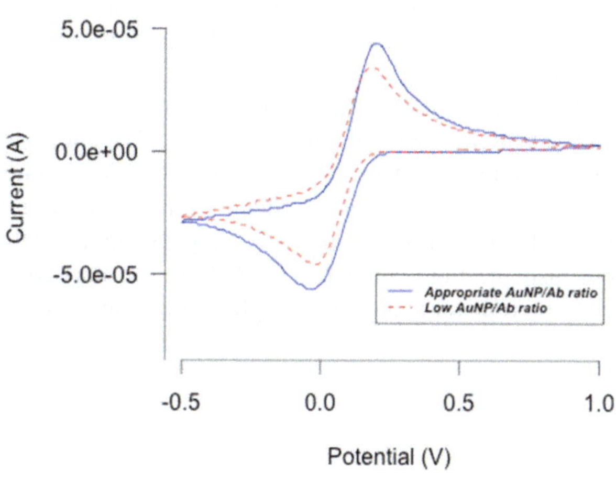

Fig. 4.8 Cyclic voltammetry results of two different concentrations of Ab-conjugated gold nanoparticles

studies have shown that the thermal optimum depends on the chemical nature of the protein, adding that hydrogen bonds are more stable at low temperatures, whereas the strength of the hydrophobic bonds increases with temperature [21]. Antigen-Ab reactions are also expected to stabilize at lower temperatures.

According to our results, however, no significant change was reported when the incubation was performed at a temperature range of 4 °C to room temperature. As a result, to simplify the process, all the processes were performed at room temperature and the optimized incubation time for this temperature was calculated to be 2.5 h, indicating saturated binding of the immunoreaction.

The pH of the solution is another factor affecting the quality of the end product. The optimal reaction rate of NHS chemistry proceeds around pH 4.5–7.2. In other words, NHS-esters hydrolyze faster in higher pH, whereas no reaction will take place at pH values lower than 3.5 [22]. As for GSH, best results are achieved by adjusting the pH at values in which the carboxylic groups are in their protonated form. This occurs when GSH is as a zwitterion (pH = 3–7) [23].

Moreover, extreme pH values induce marked conformational changes in the Ab molecule that probably may affect its complementarity with the antigen. On the other hand, since Oc, the Ab used in this work, has an isoelectric point (pI) of 4.4, at lower pH values it carries more positive charge and thus positively influences the electrostatic attraction between the oppositely charged surface carboxyl anions and the protonated amino group of the ligand (http://web.expasy.org/cgi-bin/compute_pi/pi_tool1?P02818@52-100@average). As a result, no attempts were made to change the negative pH of the sulfo-NHS solution.

4.5 Conclusions

Through the inherent interaction between AuNPs and antibody biomolecules, a novel gold nanoprobe was developed to be used in non-enzymatic electrochemical immunoassay. This one-pot method not only provides a simple method for loading high-content antibody on nanoparticles but also greatly improves the repeatability and controllability of the nanoprobe preparation. Combined with a disposable electrode, it could be used for a single/multiplexed electrochemical immunosensing method.

Generally, these electrochemical sensors have the virtue of high sensitivity and selectivity, but it requires several fabrication steps. Current nanoprobe has overcome this problem as it provides direct electron transfer between biomolecules and electrode without any mediators after being immobilized on the surface. This method of detection will not only simplify the fabrication steps with better stability but also the high-content of AuNPs in the nanoconjugate results in accelerated electron transfer and greatly amplifies the detection signal.

References

1. J. Ren-Der, C. Wei-Da, H. Meng-Hsuan, C. Fu-Hsuan, B. Jong-Shing, L. Dean-Mo, Highly electrostatically-induced detection selectivity and sensitivity for a colloidal biosensor made of chitosan nanoparticle decorated with a few bare-surfaced gold nanorods. Biosens. Bioelectron. **52**, 111–117 (2014)
2. J. Zhang, B. Ting, M. Khan, M. Pearce, Y. Yang, Z. Gao, J. Ying, Pt nanoparticle label-mediated deposition of pt catalyst for ultrasensitive electrochemical immunosensors. Biosens. Bioelectron. **26**(2), 418–423 (2010)
3. T.R. Holford, F. Davis, S.P. Higson, Recent trends in antibody based sensors. Biosens. Bioelectron. **34**(1), 12–24 (2012)
4. B. Chikkaveeraiah, A. Bhirde, N. Morgan, H. Eden, X. Chen, Electrochemical immunosensors for detection of cancer protein biomarkers. ACS Nano **6**(8), 6546–6561 (2012)
5. J. Rusling, C. Kumar, J. Gutkinde, V. Patele, Measurement of biomarker proteins for point-of-care early detection and monitoring of cancer. Analyst **135**, 2496–2511 (2010)
6. G. Laia, H. Zhanga, J. Yong, J. Yu, In situ deposition of gold nanoparticles on polydopamine functionalized silica nanosphere for ultrasensitive nonenzymatic electrochemical immunoassay. Biosens. Bioelectron. **47**, 178–183 (2013)
7. A. Yakoh, C. Pinyorospathum, W. Siangproh, O. Chailapaku, Biomedical probes based on inorganic nanoparticles for electrochemical and optical spectroscopy applications. Sensors **15**, 21427–21477 (2015)
8. F. Kurniawan, V. Tsakova, V.M. Mirsky, Analytical applications of electrodes modified by gold nanoparticles: dopamine detection. J. Nanosci. Nanotechnol. **9**(4), 2407–2412 (2009)
9. P.B. Luppa, L.J. Sokoll, D.W. Chan, Immunosensors—principles and applications to clinical chemistry. Clin. Chim. Acta **314**(1–2), 1–26 (2001)
10. M. Faraday, Experimental relations of gold (and other metals) to light. Proc. R. Soc. Lond. **147**, 145–181 (1857)
11. J. Turkevich, P. Stevenson, J. Hillier, A study of the nucleation and growth processes in the synthesis of colloidal gold. Discuss. Faraday Soc. **11**, 55–75 (1951)
12. G. Frens, Controlled nucleation for the regulation of the particle size in monodisperse gold suspensions. Nature **241**(105), 20–22 (1973)
13. H. Verma, P. Singh, R. Chavan, Gold nanoparticle: synthesis and characterization. Vet. World **72**(4), 72–77 (2014)
14. G. Hermanson, A. Mallia, P. Smith, *Immobilized Affinity Ligand Techniques* (Academic, San Diego, 1992)
15. A.M. Coto-Garcia, E.S. Sotelo-Gonzalez, M.T. Fernandez-Arguelles, R. Pereiro, J.M. Costa-Fernandez, A. Sanz-Medel, Nanoparticles as fluorescent labels for optical imaging and sensing in genomics and proteomics. Anal. Bioanal. Chem. **399**(1), 29–42 (2011)
16. A. Zuber, M. Purdey, E. Schartner, C. Forbes, B. van der Hoek, D. Giles, A. Abell, T. Monro, H. Ebendorff-Heidepriem, Detection of gold nanoparticles with different sizes using absorption and fluorescence based method. Sens. Actuators B Chem. **227**, 117–127 (2016). Available: http://www.sciencedirect.com/science/article/pii/S0925400515307711
17. G. Lai, F. Yan, H. Ju, Dual signal amplification of glucose oxidase-functionalized nanocomposites as a trace label for ultrasensitive simultaneous multiplexed electrochemical detection of tumor markers. Anal. Chem. **81**(23), 9730–9736 (2009)
18. M. Hu, J. Chen, Z. Li, L. Au, G. Hartland, X. Li, M. Marquez, Y. Xia, Gold nanostructures: engineering their plasmonic properties for biomedical applications. Chem. Soc. Rev. **35**, 1084–1094 (2006)
19. Z. Guo, Z. Liu, X. Yao, K. Zhang, X. Chen, J. Liu, X. Huang, A molecular-gap device for specific determination of mercury ions. Sci. Rep. **3**, 3115 (2013)
20. E. Day, L. Bickford, J. Slater, N. Riggall, R. Drezek, J. West, Antibody-conjugated gold-gold sulfide nanoparticles as multifunctional agents for imaging and therapy of breast cancer. Int. J. Nanomed. **5**, 445–454 (2010)

21. R. Reverberi, L. Reverberi, Factors affecting the ag-ab reaction. Blood Transfus. **5**(4), 227–240 (2007)
22. M. Fischer, Amine coupling through edc/nhs: a practical approach. Methods Mol. Biol. **627**, 55–73 (2010)
23. M. Hormozi-Nezhada, E. Seyedhosseinia, E. Robatjazia, Spectrophotometric determination of glutathione and cysteine based on aggregation of colloidal gold nanoparticles. Sci. Iran. **19**(3), 958–963 (2012)

Chapter 5
Electrochemical Chip Preparation

Abstract Since most electrochemical reactions occur in close proximity of the electrode surface, the electrodes play a crucial role in the performance of electrochemical biosensors. In other words, a biosensor is created on a solid electrode surface by chemically or electrostatically attaching bio(macro)molecules such as proteins and nucleic acids. If the surface is chemically modified, the biochemical molecules bind in a layer such that they cover the electrode surface. This layer is also known as the "recognition layer," as its molecules are specific to a target in the analyte. When the electrode is placed into a sample solution, the analyte and recognition layer interact, and an electrical signal, characteristic of the analyte, is obtained [44].

This chapter therefore explains the attempts made to develop the most appropriate electrode that fulfilled the above-mentioned characteristics for our purposes and reports its reproducibility, reusability, and stability characteristics.

5.1 Introduction

The dynamic range of concentration that a biosensor can detect is determined by the maximum loading capacity of this layer. Moreover, based on the function of chosen material on the electrode, its surface modification or its dimensions greatly influence its sensing abilities.

According to the literature, electrochemical transducers are made of different materials including carbon-based matrices such as glassy carbon, graphite, and nano-carbon, different metals such as platinum, gold, and silver, or polymers with electrical conductivity property. The desirable characteristics of the materials to be used for this purpose include:

- High versatility in shape and size
- Low cost
- Easy fabrication
- High signal-to-noise ratio
- Surface regeneration
- Suitable mechanic and conducting characteristics
- Long-term stability and lifetime

P. Khashayar, *Point-of-Care Solution for Osteoporosis Management*,
DOI 10.1007/978-3-319-55053-4_5

- Biocompatibility
- Large surface area.

When an electrode is prepared, cyclic voltammetry (CV) was used to study its properties. The electrodes were cycled twice in 0.1 mM $K_3[Fe(CN)_6]$, containing 0.01 M NACl solution from 0.0 V to + 1.2 V, at 0.1 V/s scan rate in order to get stable CV responses suitable for comparison. The same setting was used all through this work, when a CV was performed.

It is also imperative that their reproducibility, reusability, and stability are studied to guarantee a proper electrical conductivity without losing its properties over time.

- The reproducibility of the electrodes is generally tested by comparing the cyclic voltammograms of three prepared electrodes from different batches. Electrodes with less than 5% difference in the peak current are considered reproducible.
- The reusability of the prepared electrodes is assessed through evaluating current change during ten successive CV cycles.
- The stability of the electrodes is studied by comparing the baseline CV and that recorded after continuously stirring the electrodes in water for 20 min. This process was repeated four times.
- As for the storage stability, the cyclic voltammograms of the electrodes preserved at room temperature for a week are compared with that of a freshly prepared electrode using the same method.

This chapter therefore explains the attempts made to develop the most appropriate electrode that fulfilled the above-mentioned characteristics for our purposes and reports its reproducibility, reusability, and stability characteristics. It contains information published in the article from the author and her team in "Biointerface Research in Applied Chemistry" [45].

5.2 Electrode Selection

5.2.1 Carbon Electrodes

Carbon is an ideal electrode material due to its wide anodic potential range, low residual current, chemical inertness, low cost, fast response time, and suitability for various sensing and detection applications as well as ease of fabrication in different configurations and sizes. Compared with platinum or gold, carbon electrodes allow scans to more negative potentials and better anodic potential windows [1]. Carbon is observed in several variants such as glassy carbon, graphite, fullerene, graphene, and carbon nanotubes [2].

5.2.1.1 Pencil Electrodes

As the first step, pencil lead was used as the electrode because of its good electro-catalytic properties, abundance, high electrochemical reactivity, good mechanical rigidity, and low cost of the graphite [3]. Graphite, the dark gray, soft, and porous material found particularly in the pencil, has high adsorption capacity, an electrical conductivity of 2×10^5 σ (S/m) at ambient temperature, electrical resistivity of 50 $\mu.\Omega$.m, and density of 1.95–2.3 g/cm^3 [4]. It is also reported that certain polymer leads even outperform glassy carbon, basal and edge graphite, and boron-doped diamond electrodes [5].

The anisotropic nature of graphite is the result of two types of bondings acting in different crystallographic directions. In graphite, three out of four valence electrons of each carbon atom are used in bond formation with three other carbon atoms while the fourth electron of each carbon atom forms delocalized π-bonds which spread uniformly over all carbon atoms. The carbon atoms are arranged hexagonally in a planar condensed ring. The layers are stacked parallel to each other, with the atoms within the rings bonded covalently, and the layers bonded loosely through Van der Waals forces. Thus, the layers can easily slip off one to another.

Preparation Technique

Pencil lead was extracted from more than 20 brands of regular and mechanical pencils with different rigidities (H1-H6-HB-B1-B8). A wire and the electrical contact made of a conductive material such as copper was attached to one end of the lead to make the first generation electrodes. A non-conductive cylindrical shaft was press-fitted to cover the lead, protecting it from chemical reactions while inserted in the probe (Fig. 5.1). In order to facilitate the fabrication process, the

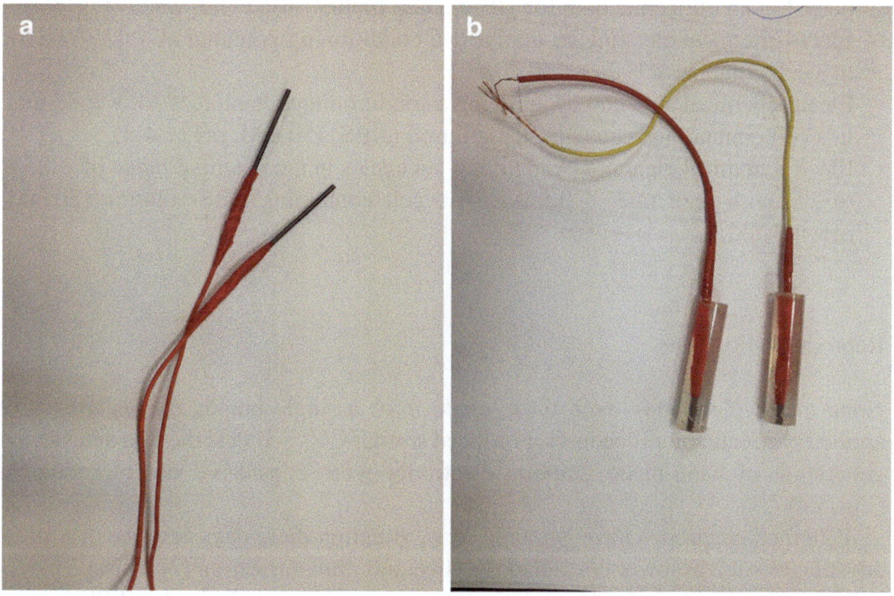

Fig. 5.1 First generation of pencil lead electrodes

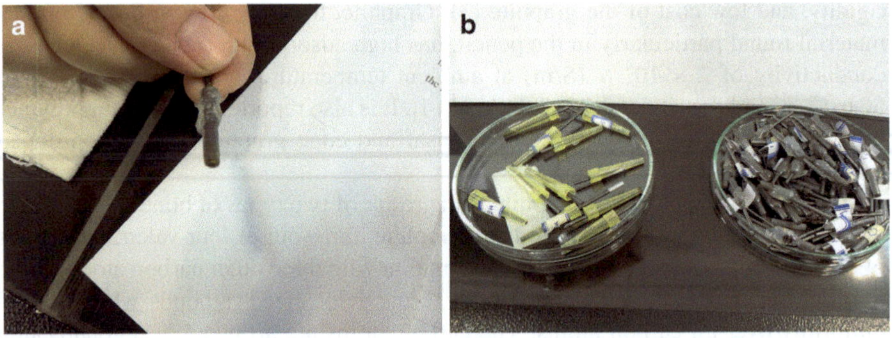

Fig. 5.2 Second generation of pencil lead electrodes

second generation electrodes were made by simply placing the pencil lead in a plastic tubing or pipette tips. Considering the conductive properties of the graphite, the pencil lead was directly connected to the system (Fig. 5.2).

A combination of pretreatment methods, listed below, were applied to eliminate any possible impurities in the lead and thus to improve the reproducibility of the electrodes:

- Pencil lead was baked in oven at 50 °C for 5 min.
- Pencil lead was polished with aqueous slurry of 0.3–0.5 μm alumina powder on a polishing cloth to a mirror finish.
- Pencil lead was placed in a cellulose thimble and then in a soxhlet extractor for 24 h of continuous reflux in methanol. Then it was washed with 0.1 M HCl, placed in an ultrasonic bath for 10 min, and finally rinsed with water (Fig. 5.3).
- Electrochemical cleaning by applying a conditioning potential of 1.4 V for 60 s, in a cell containing H_2SO_4 (50 mM).
- Electrochemical cleaning by applying a conditioning potential of 1.8 V for 5 min, in a cell containing acetate buffer solution (ABS) (50 mM, pH = 4.8).
- Electrochemical cleaning by applying ten cycles in the potential range of −0.5 V to 2 V with scan rate of 0.5 V/s, in a cell containing ABS solution (50 mM, pH = 4.8).

Reproducibility Tests

None of the electrodes, even those made from a single brand, and regardless of applied pretreatment procedures, produced similar CVs and thus the first and second generation of hand-made graphite electrodes were considered not reproducible (Fig. 5.4).

Patterned electrodes have garnered more attention these days because they offer advantages such as lower cost, disposability, and miniaturization [2] (Fig. 5.5).

Unlike the hand-made electrodes, carbon screen printed electrodes (SPEs), used as standard, produced reproducible results (Fig. 5.6).

Fig. 5.3 Pencil lead was placed in a cellulose thimble and then in a soxhlet extractor for 24 h of continuous reflux in methanol, as a pretreatment procedure

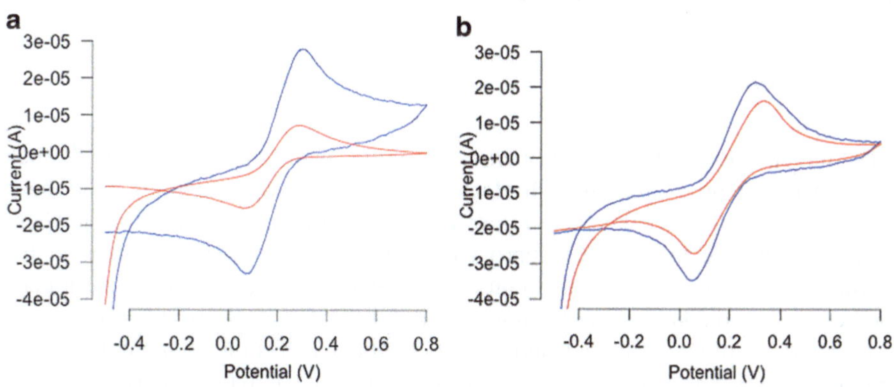

Fig. 5.4 Reproducibility studies of the hand-made graphite electrodes (**a**) first generation (**b**) second generation

Fig. 5.5 Sample carbon and gold SPEs

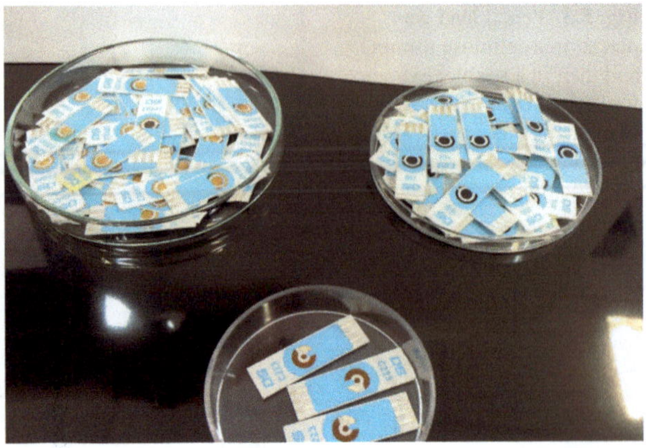

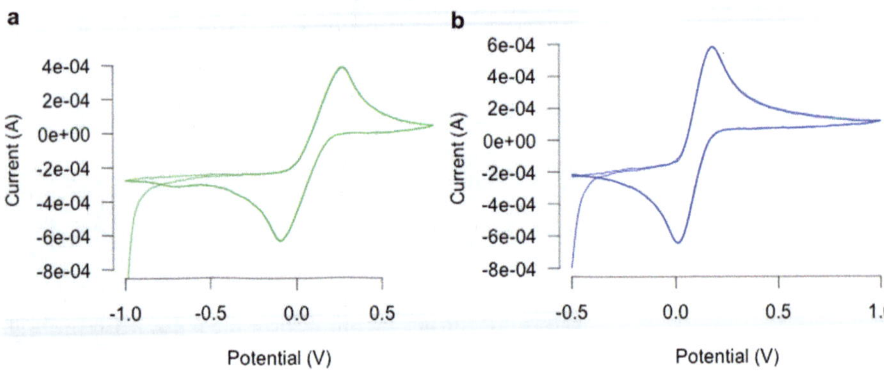

Fig. 5.6 Reproducibility results of (**a**) carbon SPEs (**b**) gold SPEs

5.2.2 *Gold Electrodes*

The unique physicochemical properties of noble metals, particularly gold, such as high mechanical strength, rich electronic, and catalytic properties, have been explored to design a new generation of electronic biosensor devices [6, 7]. These electrical and electrochemical approaches generally rely on the changes in the ohmic response of an electrical circuit or the flow of electrons arising from faradaic processes (i.e., oxidation or reduction) near the surface of an electrode, respectively. Although the direct electrical and electrochemical detection of biomolecules is possible, many have benefited from the electroactive or catalytic properties of noble metals as reporters for biosensing with unprecedented levels of sensitivity [8].

5.2.2.1 Glass-Based Gold Electrodes

Preparation Technique

White float glass substrates were thoroughly cleaned to remove organic residues. They were firstly soaked in RBS aqueous detergents, which assure wetting and penetrating actions on the residues. The mechanical effect of an ultrasonic bath permitted the removal and dispersion of the residues. The cleaned substrates were finally rinsed by soaking them in isopropanol (IPA) and DI water, and dried. Next, a 50-nm titanium tungsten [TiW (10%/90%)] adhesion enhancement layer, followed by 500-nm copper, 50-nm titanium tungsten [TiW (10%/90%)], and 50-nm gold films were sputtered within one vacuum cycle of an Alcatel SCM600 system.

Later, however, the Cu layer was eliminated as it was damaged by H_2SO_4 used for cleaning. The thickness of the Au layer was also increased to improve conductivity. In the final design, a 50-nm titanium tungsten [TiW (10%/90%)] adhesion enhancement layer and a 100-nm gold film were sputtered on the cleaned glass substrate. The metal films were subsequently patterned by photolithography and subsequent wet etching.

The electrode design was fabricated using common computer aided drawing (CAD) software and printed onto a transparent foil by the JD photo-tools in the UK. The design followed several considerations:

- The primary consideration for the relative location of the three electrodes in a cell is the minimization of the solution resistance. This was accomplished by keeping the three electrodes as close together as feasible, while not interfering with the current paths between one another.
- The second important consideration involves the shape and size of the auxiliary electrode (AUX) relative to the working electrode (WE). The Aux should be at least as large in area as the WE, and positioned symmetrically with respect to the WE so that the current density and potential experienced along its entire length is constant.
- Close proximity of the reference (Ref) and WE electrodes is important.

To this end, positive photoresist S1818 was used as masking layer to wet etch both metal films. A 1.8 μm thick coating of S1818 was spin-coated at 4000 rpm for 60 s and soft baked on a hot plate for 2 min at 90 °C. Next, the photoresist was UV exposed through the foil mask with the desired pattern for 7.5 s at 13 mW/cm^2, developed for 30 s in 50% Microposit developer, and hard baked for 30 min in a convection oven at 120 °C. The gold and titanium tungsten films were then wet etched using a solution of iodine (I_2) and potassium iodide (KI) [1:4 (1 g I_2, 4 g KI, 73 ml H_2O)] at room temperature, and a 30% H_2O_2 solution at 55 °C, respectively. Finally, acetone was used to strip the remaining photoresist from the electrode surface.

Since the connection lines between the electrodes and their connections points were all made of gold, SU-8 was used as an insulation layer to cover the whole

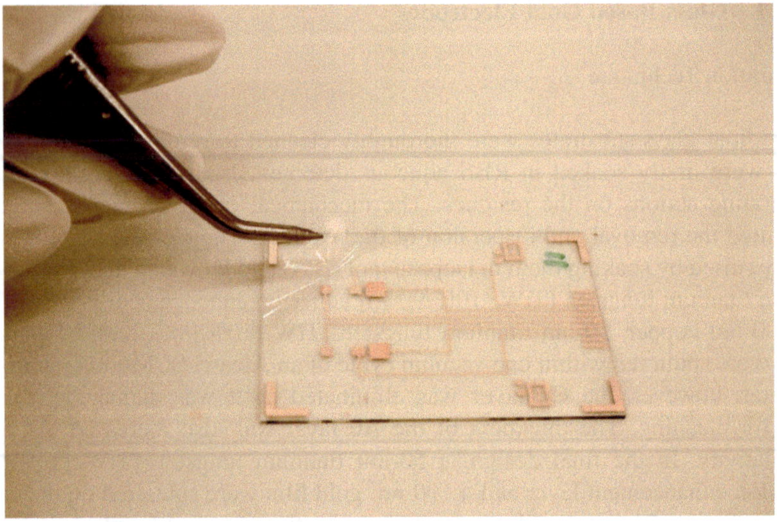

Fig. 5.7 SU-8 delamination in the absence of TI PRIME

surface, exposing just the electrodes to the electrolyte solutions. SU-8 was used because it was biocompatible and spectrophotometry results showed its negligible protein adsorption rate[1] [9].

In this regard, TI PRIME was first spin-coated on the substrate to improve the adhesion of SU-8 to the glass substrate. In the absence of TI PRIME, the SU-8 layer was delaminated easily (Fig. 5.7). 10-μm thick layer of the negative-tone photoresist SU-8 3010 was then spin-coated on the substrate. After the soft bake (65 °C for 1 min, followed by 95 °C for 3 min), the substrate was allowed to cool down to room temperature. Next, the coating was exposed to UV radiation (300 mJ/cm2) through a glass mask, using a double side mask aligner (EVG 620; Austria). A long pass filter was used to eliminate UV radiation below 350 nm. After the post-exposure bake (65 °C for 1 min, followed by 95 °C for 2 min), the SU-8 was developed (90 s in SU-8 developer), followed by an IPA rinse. Finally, hard bake was performed (120 °C for 90 min) to further cross-link the SU-8, after which the substrates were diced (Fig. 5.8).

In the final design, the electrochemical chip consisted of six gold electrodes arranged in two by three electrode arrays on a glass substrate. The glass was $25.5 \times 85.5\,mm^2$, leaving a $25.5 * 10\,mm^2$ uncovered area for the contact pads to be connected to the reader using a printed circuit board (PCB). The PCB with electrical connection leads and lines to the electrodes were used to connect the array to the potentiostat. Each gold working electrode ($1.5 \times 1.5\,mm^2$) was placed between one gold pseudo-reference ($1.4 \times 1.4\,mm^2$) and one gold counter electrode ($2.5 \times 2.5\,mm^2$) (Fig. 5.9).

The gold electrodes were dried using N_2 and kept at 4 °C prior to use.

[1]The process is explained in detail in Chap. 7.

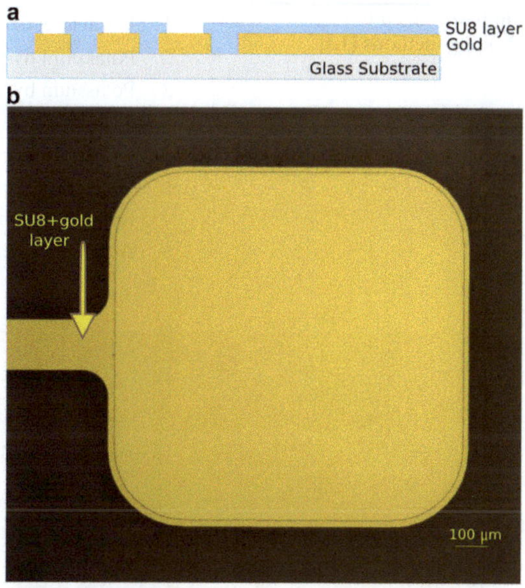

Fig. 5.8 (**a**) Schematic illustration of the cross-section of an electrochemical chip (**b**) Optical microscope image of one of the electrodes on the electrochemical chip with SU-8 layer on its edges and connection line

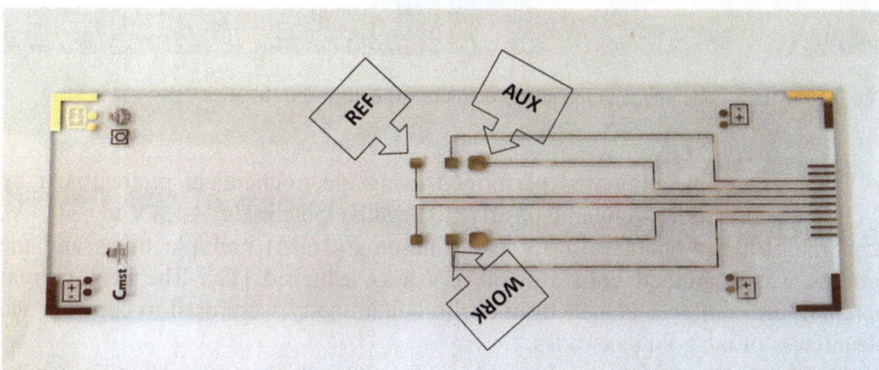

Fig. 5.9 Structure of the electrochemical chip manifold

Cleaning Technique

Upon exposure to a non-cleanroom laboratory environment, the gold surface is subject to numerous ambient contaminants (AuO, hydrocarbons, etc.), affecting binding kinetics of thiols in addition to electrochemical effects. As a result, gold must be cleaned immediately before it is chemically modified for electrochemical uses (Table 5.1).

Before storing the electrodes, they were etched by means of radicals, formed in the gas mixture plasma, chemically reacting on the surface of the substrate, and physically by means of ions bombarding the surface in the Reactive Ion Etching (RIE) machine. The goal was to remove possible contaminating material deposited on the electrode surface (Fig. 5.10).

Table 5.1 Common gold cleaning methods [10]

1	Ultraviolet ozone cleaning
2	Potassium hydroxide and hydrogen peroxide cleaning
3	Potassium hydroxide potential sweep
4	Sulfuric acid and hydrogen peroxide cleaning
5	Sulfuric acid potential cycling
6	Hydrochloric acid potential cycling
7	Reducing agent solutions cleaning
8	Aqua regia cleaning

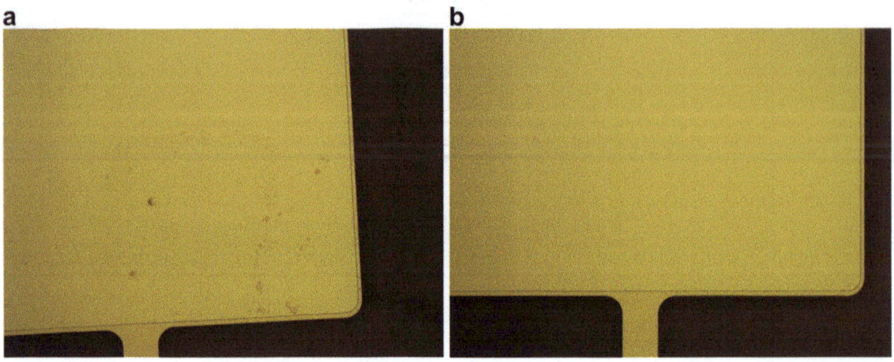

Fig. 5.10 Microscope image of the gold electrodes (**a**) before and (**b**) after RIE

Upon usage, cleaning was performed using electrochemical pretreatment by cycling the electrode from large positive to negative potentials (+1.4 V to − 0.4 V, scan rate 100 mV/s) in sulfuric acid solution (50 mM) multiple times and the process was continued until a stable CV was achieved [11]. The peak-current potential differences obtained from cyclic voltammetry were used to compare the cleanliness of the gold electrodes.

At this potential range, gold oxide is formed, which upon reduction can be stripped away, together with other surface contaminants and possible remnants of the photoresist on the electrode surface. We found that after approximately 20 cycles, the recorded CVs were reproducible and featured characteristics of a clean gold surface became visible. Further CVs showed a remarkable repeatability of the as mentioned treated Au working electrode.

Performance Analysis

Considering the fact that all the working, auxiliary, and reference electrodes all together are present on the electrochemical chip, the accuracy of the system was checked as follows.

First, as the voltage that is applied to the working electrode is controlled with respect to the reference electrode, the stability of the reference electrode was measured based on the consistency between cycles of the potential at which the peaks occurred, compared with an external reference electrode was used. The average, standard deviation, and relative standard deviation (RSD) were calculated for the peak potentials within nine successive CV cycles of one measurement. A less than 5% variation was noted between cycles, suggesting not only the reference electrode was stable over the time period in which the measurement was collected, but also the difference between using an on-chip and external reference electrode was ignorable.

However, considering the slightly different potential and offsets between the working and reference electrodes, commonly reported in sensors using quasi-reference electrodes such as ours, we averaged all the electrochemical plots for each of the studied electrodes [12].

Moreover, as the current in the system passes between the working and counter electrodes, the stability of these electrodes was determined by examining the variation in maximum current in each peak and comparing the results to that reported when using an external counter electrode. Similarly, a less than 5% reported variation suggested the difference between using an on-chip and external counter electrode to be ignorable.

Reproducibility Tests

The as prepared gold electrodes were reported to be reproducible and stable during successive CV analyses (Fig. 5.11). Again the results were compared with gold SPE as a standard (Fig. 5.5).

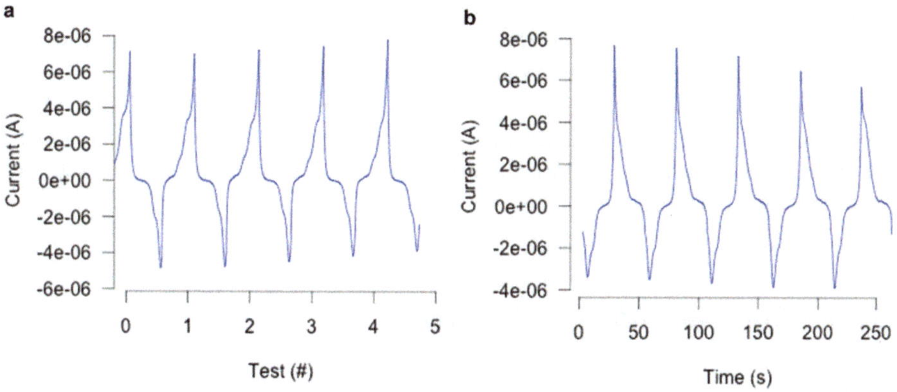

Fig. 5.11 CV analyses showed the gold electrodes were (**a**) reproducible and (**b**) stable during successive CVs

5.3 Surface Modification of Gold Electrodes with AuNPs

The purpose of electrode modification with nanostructures is to improve the sensitivity and selectivity of the biosensor. The higher surface-to-volume ratio of nanostructure enhances the electrical properties, which makes the electrode susceptible to external influences. Since the dimension of nanostructure becomes similar to the size of the target biomolecules, measurement sensitivity will increase, and this will enhance higher capture efficiency [13].

In other words, these electrodes enhance the analytical performance of the sensor through providing a suitable microenvironment for biomolecule immobilization, their effects on increasing the surface area of the exposed gold, and facilitated electron transfer between the immobilized proteins and electrode surfaces. As mentioned earlier, AuNPs are the most common nanostructures used in this regard.

According to the existing studies, AuNP deposition results in three-dimensional Au deposits, which have a higher desire to adsorb specific functional groups [14, 15]. The size, structure, and distribution of these structures, however, have a strong effect on the performance of the system and thus should be well controlled.

Diverse strategies have been reported to coat various working electrodes such as platinum, carbon, and gold with an AuNP layer [16, 17]. These coating techniques include layer-by-layer (LBL) deposition of AuNPs, self-assembled monolayer (SAM) development, sol-gel technique, impregnation, co-precipitation, metal organic-chemical vapor deposition, incipient wetness, electrodeposition, and dip-coating [18, 19].

Some of them, such as the fabrication of metal nanoparticles layers using electron-beam lithography (EBL) or nanotemplating, require expensive equipment or special templates and are thus less commonly used [20]. Many studies have also reported that the use of electrodeposition techniques is preferred, compared to physical vapor deposition (PVD) methods, because of their relatively low-cost and potentially higher deposition rate and film thickness [21, 22].

The aim of this section is to compare the most common existing methods for depositing gold nanoparticles on gold electrodes for electrochemical sensing purposes and present an optimized, adopted protocol to increase the surface area and analytical performance.

5.3.1 Gold Deposition Techniques

In this work, deposition of the gold nanostructures was performed through different methods. Some of them are simple modification of methods adopted from literature, whereas others are replications of techniques previously used for other purposes.

5.3.1.1 Method 1: Layer-by-Layer Deposition of AuNP

Previous studies have suggested that the interactions in the spherical electron density around the gold atoms are responsible for the formation of gold–gold bond in a reliable way [23]. Moreover, some studies have shown the high binding affinity of the terminal functional groups on the periphery of the ligand shell of the gold nanoparticles and the gold electrode surface [24]. They have shown the occurrence of the ligand replacement reactions when this binding affinity exceeds that of the ligand to the particle. As a result, this kind of deposition could be performed in different ways:

(a) Multilayer AuNP-electrodes—2 μl of the prepared aqueous solution of AuNP was deposited on the gold electrodes,[2] and was subsequently dried at room temperature. This process was repeated three times. The electrodes were rinsed with deionized water and dried in a stream of nitrogen gas before the next layer deposition.
(b) In order to improve the results, the layer-by-layer (LBL) deposition technique was applied [25]. The LBL method is based on the alternate deposition of oppositely charged (bio-) polymers, in our case poly(diallyldimethylammonium chloride) (PDDA) [26]. PDDA is commonly used as a reducing and stabilizing agent to avoid AuNP aggregation. When deposited through LBL, the water-soluble, quaternary ammonium, cationic polyelectrolyte enables control over the total AuNP/polymer thickness. The negatively charged substrate also improves attachment of positive compounds.
(c) The same process as mentioned in method 1a was repeated using PDDA-protected AuNPs. The electrode was rinsed with deionized water and dried in a stream of nitrogen gas before the next layer deposition.
(d) According to another method in literature, the previous process was repeated on a GSH-modified gold electrode [27].
(e) The gold electrode was immersed in PDDA solution (20% aq solution in 0.5 M NaCl) for 10–30 min. After drying, 2 μl of the prepared aqueous solution of GSH-modified AuNP was deposited on the gold electrode and dried at room temperature [28]. This process was repeated three times. The electrode was rinsed with deionized water and dried in a stream of nitrogen gas before the next layer deposition.
(f) In another attempt, the PDDA-coated substrate was immersed into $HAuCl_4$ solution (10 mg/mL) for 15 min [29]. Thereafter, the electrode was rinsed with deionized water and dried under N_2 before the next layer deposition. Subsequently, the electrode was placed in 0.1 M KNO_3 solution and −0.4 V potential was applied for 60 s. The process was repeated three times.

[2]Preparation of AuNP is explained in Chap. 4.

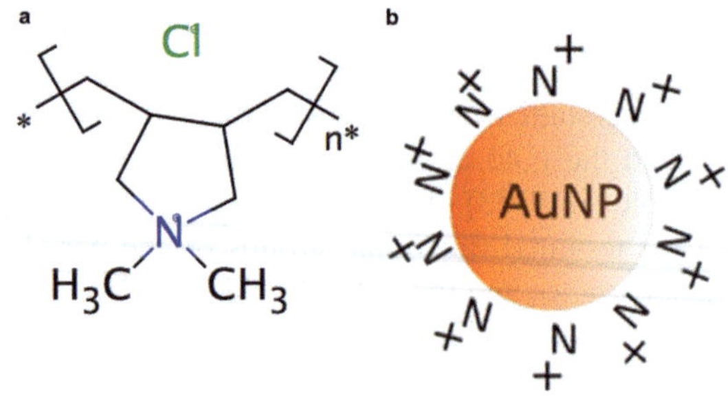

Fig. 5.12 (**a**) PDDA structure, (**b**) schematic of PDDA-protected AuNP

Preparation of PDDA-Protected AuNPs

Hereof 250 μL of PDDA solution, 40 mL water, 200 μL 0.5 M NaOH, and 100 μL $HAuCl_4$ (10 mg/mL) were mixed and subsequently heated to 100 °C with gentle stirring until the color changed to red. The UV-Vis spectrum showed that the band with a maximum λ at 520 nm in the colloid gold solution was shifted rightward to $\lambda_{max} = 523$ nm (Fig. 5.12).

Preparation of GSH-Modified AuNPs

GSH is a tripeptide biomolecule (–Glu-Cys-Gly), containing a –SH group, which can be easily adsorbed onto the AuNP surface. There are functional groups ($-NH_2$, –COOH) which have strong affinity to metal ions in GSH (coordination bonds). A dense layer of GSH-AuNP enlarges the electroactive surface area and provides carboxylic groups on the surface of AuNP to covalently bind a dense layer of antibody/protein for detection. Best results are reported when GSH is available in zwitterion (pH = 3–7) as pH affects the -OH, -COOH, $-NH_2$ groups in the protein and the aggregation of GSH benefits from H-bonding, which is pH dependent (best results were reported in the acidic environment). The aggregation rate decreases with an increase in NP concentration (Figs. 5.13 and 5.14).

A stock solution of glutathione (10 mM) was prepared. Dilute HCl solution (10 mM, 100 μL) was added to the gold aqueous solution (3 mL) to lower the pH (~4) [30]. An aliquot of GSH solution (10 mM, 30 μL) was added to the acidic gold aqueous solution, until the color changed from deep red to blue, indicating the formation of aggregates amongst AuNPs.

5.3.1.2 Method 2: Self-Assembled Monolayer Development

SAM could be achieved after deposition of cysteamine, mercaptocarboxylic acid, 3,3-dithiodipropionic acid, 2-mercaptoethanesulfonate, 4-aminothiophenol, triazole, and ultrafin polypyrrole tetradecyl sulfate as well as the composite of 11-mercaptoundecanoic acid-polyethylene glycol and many other options on gold

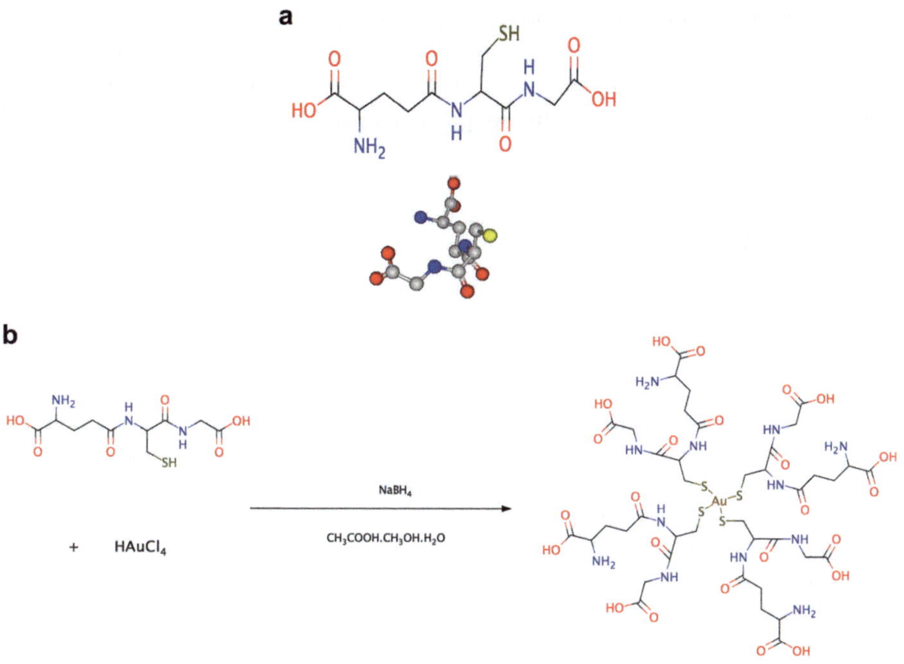

Fig. 5.13 (**a**) GSH structure, (**b**) schematic of GSH-decorated AuNP

Fig. 5.14 UV-Vis spectra of GSH-decorated AuNP solution

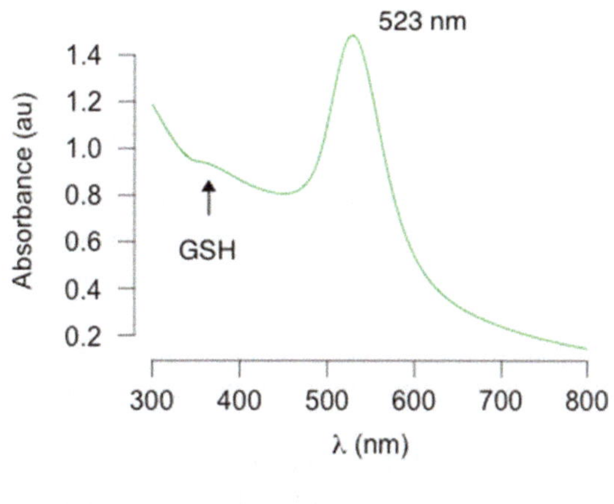

electrodes [31, 32]. It has been proved that the three-dimensional monolayers of S-containing organic compounds facilitate the electron transfer in the redox reaction occurring at the electrode–solution interface, ensuring an easier access of reagents to the electrode surface [33].

A thiol molecule consists of (1) a sulfur head group, which forms a strong, covalent bond with the gold electrode, (2) a hydrocarbon chain of variable length, which stabilizes the SAM through van der Waals interactions, and (3) a terminal group, which has different functionalities. In our case, the -SH-terminated thiols of the dithiol used made the binding of metallic ions of the gold electrode and gold nanoparticles to the SAM possible [34]. Studies have shown that the adsorption time depends on the nature of the molecule, adding that dithiol deposition from solution needs less time [35]. This is while long chain alkanethiols need 2–12 h, compared with short chain alkanethiols or thiols with certain end groups different from -CH_3 that need at least 24 h to form a well-ordered SAM.

When using dithiols, the SAM is developed using the following reactions [Eqs. (5.1), (5.2)]:

$$HS(CH_2)_nSH + Au \rightarrow (HS(CH_2)_nSH)_{phy}Au \tag{5.1}$$

$$(HS(CH_2)_nSH)_{phy}Au + Au \rightarrow Au - S(CH_2)_nSAu + H_2 \tag{5.2}$$

Moreover, dithiols provide a more stable SAM compared with alkanethiols because of the formation of two thiolate bonds per molecule in dithiols and the higher negativity of their reductive peak potentials [19, 36]. On the other hand, these bonds can hinder the transition from the lying down to the standing up configuration, particularly in dithiols with short chain length, resulting in smaller chain–chain interactions, smaller gain in energy, and thus higher stability.

Taking this all into account, we decided to use 1, 8-octanedithiol. In this regard, SAM was prepared by immersing gold electrodes into 1 mM 1, 8-octanedithiol in ethanol for 24 h in a dark environment at room temperature [37]. The modified electrodes were then taken out, soaked in distilled water, and subsequently dried in a stream of nitrogen gas.

5.3.1.3 Method 3: Electrodeposition

Electrochemical deposition is an important technique as it not only involves simple instrumentation and operation but also is versatile and cost-effective. It is generally performed through application of an electrical field to a solution containing the ions of the target material. Electrically enhanced diffusion of the reactive ions towards the electrode followed by an electron transfer from or to the ion results in oxidative or reductive deposition of the desired material on the electrode, respectively. Figure 5.15 shows a voltammogram of an electrodeposition process using CV. The reverse scan of the voltammogram in these experiments exhibits the reduction of Au(III) to Au(0), inducing the deposition of AuNPs onto the electrode surface according to the following reaction [Eq. (5.3)]:

$$H^+[AuCl_4]^-_{aq} + 3e^- \rightarrow Au^0_s + 4Cl^-_{aq} + H^+ \tag{5.3}$$

Different protocols have been proposed for gold electrodeposition on gold electrodes. Using them, various morphologies, including rod-like and dendritic gold nanoparticles, have been obtained in a one-step process. In this work, electrodeposition was performed in an aqueous solution of 1.0 mM $HAuCl_4$ and 0.5 M H_2SO_4 in

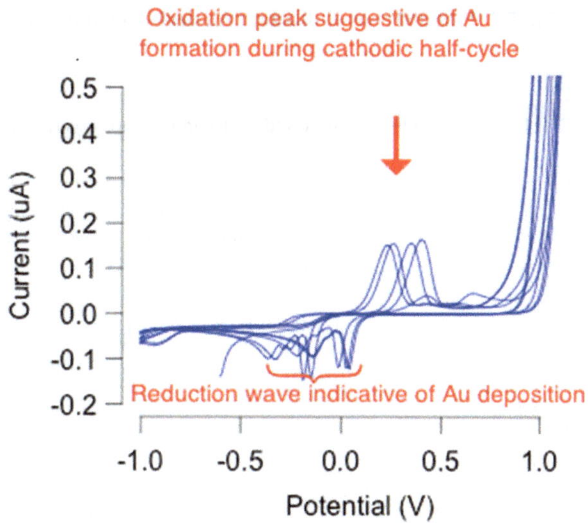

Fig. 5.15 Voltammograms of the ECV process in 1.0 mM $HAuCl_4$ and 0.5 M H_2SO_4 in the presence of 0.1 mM NaCl

the presence of 0.1 mM NaCl as the supporting electrolyte. For the electrodeposition step, external AgCl reference electrode and Pt. gauze (Alfa Aesar Gmbh & Co, Germany) as auxiliary electrode were used.

(a) Cyclic voltammetry: Electrodeposition was performed by CV scanning the potential from −0.5 V to + 0.5 V, +1.0 V, and + 1.5 V at a scan rate of 0.1 V/s for a given number of scans [38]. The process was repeated until the CV became reproducibly stable. The sweeping potential, the scan rate, and the number of cycles needed were then optimized based on the results. This method is called ECV all throughout this book.

    There was a broad reduction wave at 0.0 V suggestive of AuNP deposition. As the number of cycles increased, there was a sharp oxidation peak at 1.0 V. Throughout the electrodeposition process, the current read at this peak increased during each cycle, which is consistent with thermodynamics that predicts an easier growth of previously formed AuNPs than a nucleation of new AuNPs on the electrode. In addition, the peak area of the reduction of gold oxide suggests that a large amount of gold nanoparticles have been deposited on the surface of the electrode. After 60 repeated scans, no more change was noted in peak height or shape.

(b) Pulsed electrodeposition: The potential was stepped from −0.5 V to + 0.5 V, +1.0 V, and + 1.5 V with a 5 s deposition time [39]. NP-like gold islands were generated on the electrode; the size and density of the islands increased with increasing number of pulsed electrodeposition cycles.

Table 5.2 Pros and cons of different Au-deposition protocols applied in this work

	Pros	Cons
LBL deposition	Simple and rapid	Unstable, not reproducible
SAM development	Strong bonding to underneath layer, stable and reproducible	Time consuming, low sensing performance enhancement
Electrodeposition	Acceptable bonding to underneath layer, acceptable sensing performance enhancement, stable and reproducible	Time consuming, no. of cycles depend on the electrode characteristics

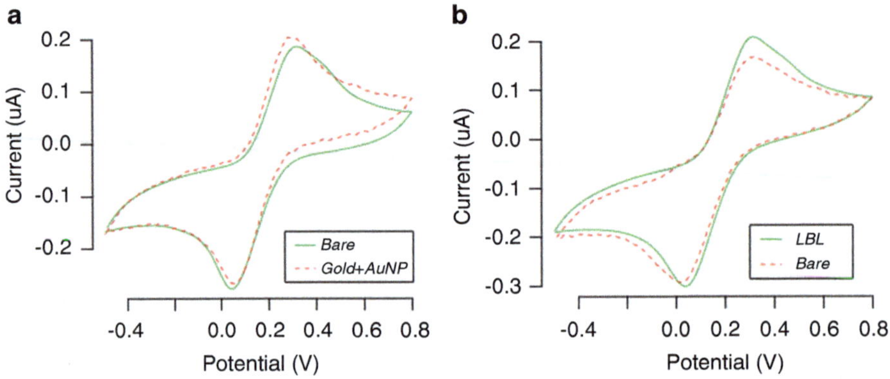

Fig. 5.16 Comparison of voltammograms of bare electrodes and electrodes developed by (**a**) method 1a and (**b**) method 1b in 0.1 mM $K_3[Fe(CN)_6]$

5.3.2 *Performance Analysis*

The comparison between the different protocols applied in this work is outlined in Table 5.2. Previous studies had suggested that not only the stability of the AuNP layer but also the density and size of the NPs are of great importance in the sensors' analytical performance, stressing that high density of rather small, spherical-shaped NPs are associated with best responses [40].

The results of the CV analysis showed that adding a layer of AuNP using method 1a did not affect the results significantly (Fig. 5.16).

While adding an AuNP-layer is believed to provide a larger surface area for the subsequent layers to bind, the unstable attachment between AuNPs and the electrode might be the reason behind the absence of any change in the results. Similar results were noted for electrodes modified using method 1c. Despite the stronger attachment of PDDA-protected AuNPs to the underlying layers, their non-conductive nature was unfavorable in biosensing processes. This was shown based on a slight decrease in peak current in CV results, suggestive of reduced conductivity of the surface (Fig. 5.17).

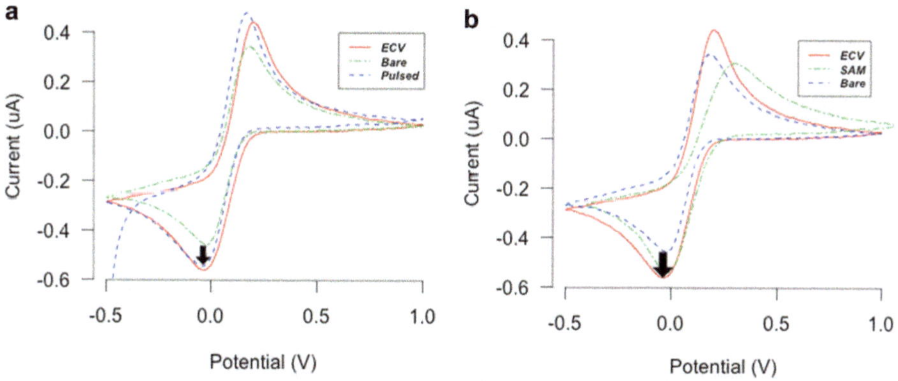

Fig. 5.17 Comparison of voltammograms of bare electrodes and electrodes developed by each of the above-mentioned techniques in 0.1 mM $K_3[Fe(CN)_6]$

Fig. 5.18 Voltammograms of electrodes prepared using method 2 in 0.1 mM $K_3[Fe(CN)_6]$ over time

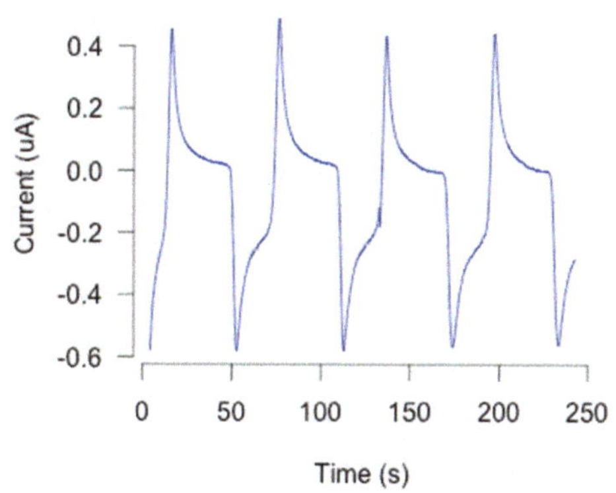

CV analysis of the electrodeposited electrodes (method 3a) showed higher current compared with methods 1 and 2, indicating that gold nanoparticles have been successfully attached to the surface of the electrode and thus improving the conductivity and surface area.

Figure 5.17 shows a considerable increase in the electrical response after SAM development. In comparison to LBL, the SAM technique provided a more stable nanoparticle layer (less than 5% variability in peak currents after ten CVs) (Fig. 5.18).

CV analysis of the electrodeposited electrodes showed higher current compared with previous methods, causing improved electrode kinetics and a reduction in the oxidation potential (thermodynamically feasible reaction). This is attributed to the larger surface area of the modified electrodes and synergistic effects of electrical conductivity and electroactivity of AuNP.

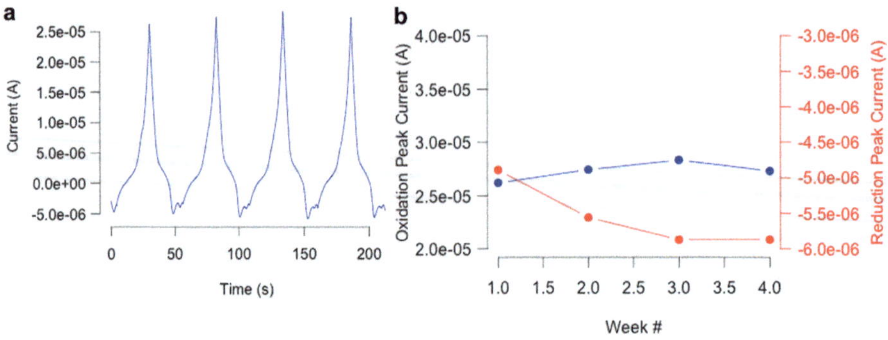

Fig. 5.19 Comparison of voltammograms of electrodes prepared using method 3a in 0.1 mM $K_3[Fe(CN)_6]$ (**a**) during successive CV cycles (**b**) over several weeks

5.3.3 Reproducibility, Reusability, and Stability Tests

The reusability of biosensors is also of great importance. The electrodes modified through LBL deposition were unstable and not reproducible. On the other hand, adding PDDA to the process did not improve the bonding of gold nanoparticles to the electrode surface considerably and thus no significant improvement was noted in the stability results (more than 5% variability in peak currents of ten successive CVs). Similar results were noted after surface modification using methods 1d and 1e. This is while the nanostructures were firmly immobilized on the gold electrode prepared using SAM and ECV techniques, resulting in an acceptable stability and reproducibility.

Similar to electrodes prepared using SAM techniques, there was no noticeable change of currents during the successive CV cycles (less than 5% variability in peak currents), which implies that the nanostructures, prepared using ECV technique, are firmly immobilized on the gold electrode (Fig. 5.19).

Our results suggested that while AuNPs deposited using both SAM and electrodeposition techniques provided a stable layer, the electrodes prepared using electrodeposition showed higher current and thus electron transfer rate and analytical performance. This is while many of the previous studies had suggested SAM techniques as an effective way for stable protein attachment [41] (Table 5.2). It has been evidenced that several factors including the packing density as well as chain length of the SAM play an important role in surface coverage and electron transfer rate. In other words, if SAM is too densely packed, similar to our results, it would exhibit a blocking behavior due to its high resistance, a process which would decrease electron transfer rate and subsequently delay the sensing process [42].

The study of the storage stability of the ECV-modified electrodes showed a slight current change (less than 5% variability), showing their long-time stability (Fig. 5.19). This, however, was not true for the electrodes prepared using pulsed electrodeposition techniques (Fig. 5.20).

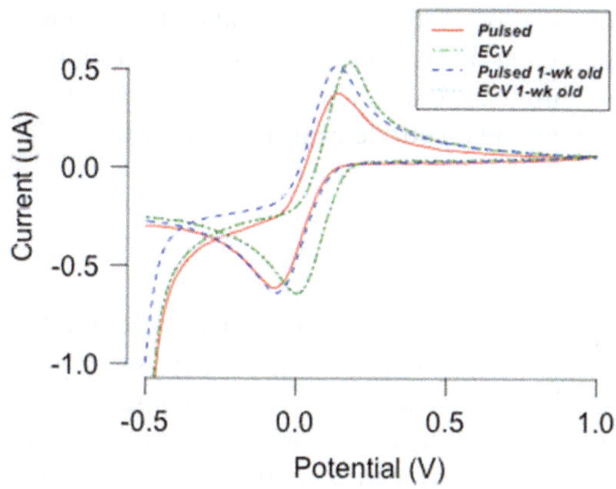

Fig. 5.20 Comparison of voltammograms of electrodes prepared using method 3a and 3b in 0.1 mM $K_3[Fe(CN)_6]$

5.4 Characteristics of AuNP-Coated Gold Electrodes Using ECV

Considering the results, the electrodeposition process of AuNP on gold electrode using ECV method was optimized and the focus of the remaining part of this section will be on the characterization of the electrodes modified using this technique.

5.4.1 Electrochemical Behavior

In this work, AuNPs were electrodeposited by ECV starting from $HAuCl_4$ (method 3a). The successful AuNP formation on the treated electrodes was confirmed by showing a typical CV of gold over the potential range of −0.5 V to +1.2 V in 0.5 M H_2SO_4 solution (Fig. 5.15). This voltammogram shows a single sharp reduction peak at +0.2 V and oxidation peaks at +1.2 V.

5.4.2 Surface Roughness

The higher current noted in further voltammograms confirmed that gold nanoparticles immobilized on the gold interface using this method increased the electrode surface area and facilitated the charge transfer between the AuNPs and the gold electrode (Fig. 5.17).

From the voltammetric responses obtained in $K_3[Fe(CN)_6]$ information regarding the real electrode surface area and its roughness were extrapolated. The diffusion coefficient of the mediator was calculated by performing CV with three different scan rates (0.1, 0.3, 0.5 V/s). The resulting currents were then plotted against the square root of the scan rate and the obtained slope was equal to the diffusion coefficient. In our case, the diffusion coefficient was calculated to be $5e-08\,\text{cm}^2/\text{s}$.

Chronocoulometry was then performed to measure the electrical charge passing through the electrode as a function of time. The analysis of the chronocoulometric data was then performed based on the Anson equation [Eq. (5.4)]:

$$Q = 2.n.F.A.C.D^{(1/2)}.\pi^{(-1/2)}.t^{(1/2)} \tag{5.4}$$

Herein, Q represents the charge (coulombs), n the number of electrons transferred, A the real electrochemical surface area of the electrode (cm^2), F the Faraday's constant (96,485 coulombs/mol), C the concentration of the mediator (mol/cm^3), D the diffusion coefficient of the mediator (cm^2/s), and t time (s).

The slope (a) of the as expected linear plot between Q and square root of time was used to calculate the real electrochemical surface area of the electrode [Eq. (5.5)]:

$$A = \frac{a}{2.n.F.C.D^{(1/2)}.\pi^{(-1/2)}} \tag{5.5}$$

As expected, the real electrochemical surface area of the gold electrodes were calculated to be $0.0784\,\text{cm}^2$, which is 2.3 times greater than the area before electrodeposition (geometric area = $0.0225\,\text{cm}^2$). Based on these results, the roughness factor (ratio of electrochemical surface to geometrical surface) was calculated to be 3.49 (compared to 1.53 in bare sputter deposited Au electrodes before electrodeposition). These results were further confirmed with SEM and AFM.

5.4.2.1 Electroless Gold Plating vs. Electrodeposition

In another attempt, a group of electrodes were subjected to electroless gold plating. The technique increased the roughness of the surface but failed to improve the sensing properties of the electrodes. This could be explained due to the fact that the deposition of AuNPs on the surface improves the biocompatibility of the surface and results in a 3D structure with higher desire to adsorb certain functional groups. It also improves direct electron transfer from the redox protein to the electrode surface.

WYKO non-contact optical profiler confirms a significant increase in the roughness of the surface following electroless gold plating (Figs. 5.21 and 5.22).

5.4.3 *Surface Modification*

In order to further characterize the morphology of the nanostructured gold film deposits, SEM (Fig. 5.23) and AFM (Fig. 5.24) analysis was performed. These images also confirmed the considerable increase in the roughness of the surface

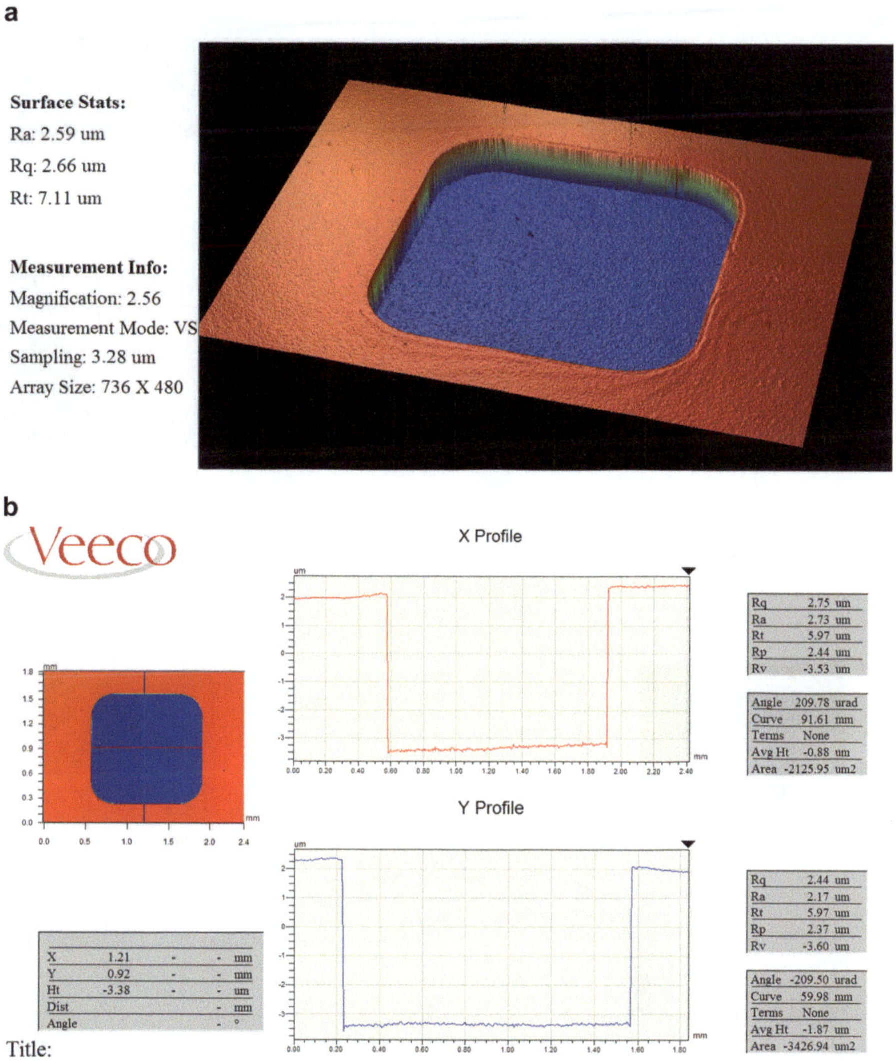

Fig. 5.21 Representative (**a**) three-dimensional image and (**b**) surface profile of bare electrode obtained using Wyko profilometry

after electrodeposition and increasing the number of electrodeposition cycles was associated with growing density of deposited nanoparticles.

Figure 5.25 shows typical nanopyramidal structures obtained from different electrodeposition conditions. An increased number of deposition cycles were associated with a significant rise in the density of these nanostructures.

STEM images revealed the formation of nanoclusters of gold nanoparticles with the mean size of 20 nm after the ECV process (Fig. 5.26). This is while as mentioned earlier, the electrochemical characteristics of AuNPs are reported to be strongly size-dependent, and due to geometrical considerations, AuNPs of $\sim$ 20 nm lower

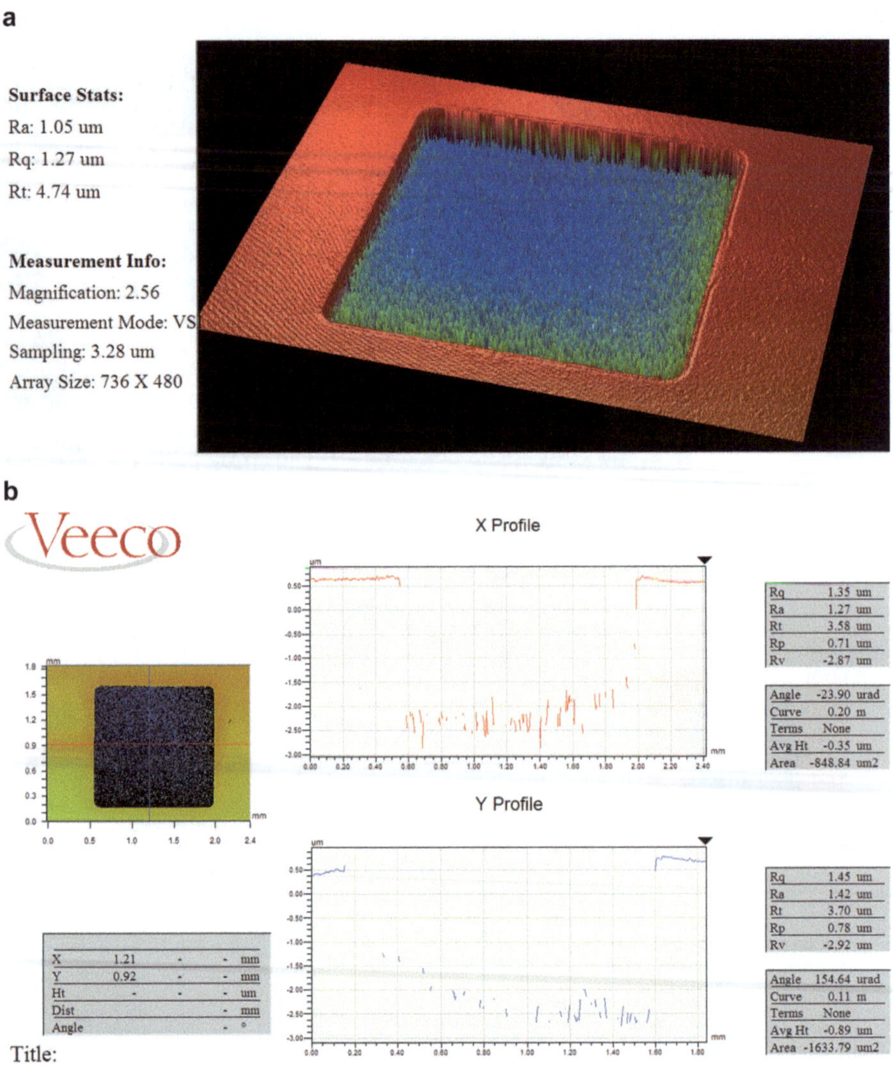

Fig. 5.22 Representative (**a**) three-dimensional image and (**b**) surface profile of electroplated electrode obtained using Wyko profilometry

the probability of proteins finding sterically suitable sites on gold [43]. The quantum effects, originated by d-band electrons of the surface shifted towards the Fermi-level, enables AuNPs to interact with electrocatalytic reactions.

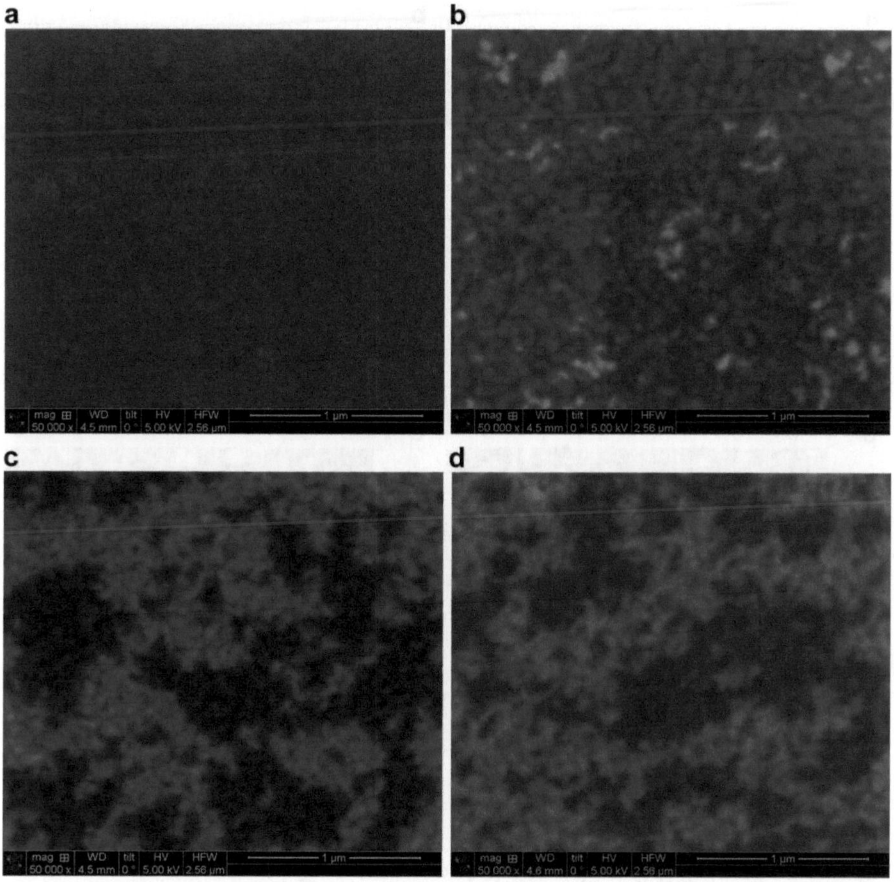

Fig. 5.23 SEM micrographs of (a) bare gold electrode, and gold electrodes after ECV with (b) 5 cycles, (c) 10 cycles, and (d) 30 cycles. (For (a), (b), (c), (d), scale = 1.0 μm)

5.4.4 Reproducibility

In order to measure the reproducibility of the as-prepared electrodes, repeatability studies were performed. In this regard, peak current in similar CVs performed on a group of electrodes were compared (Fig. 5.27). According to the results, the calculated RSD was (coefficient of variation = 5.3%), suggesting high likelihood of producing similar electrodes.

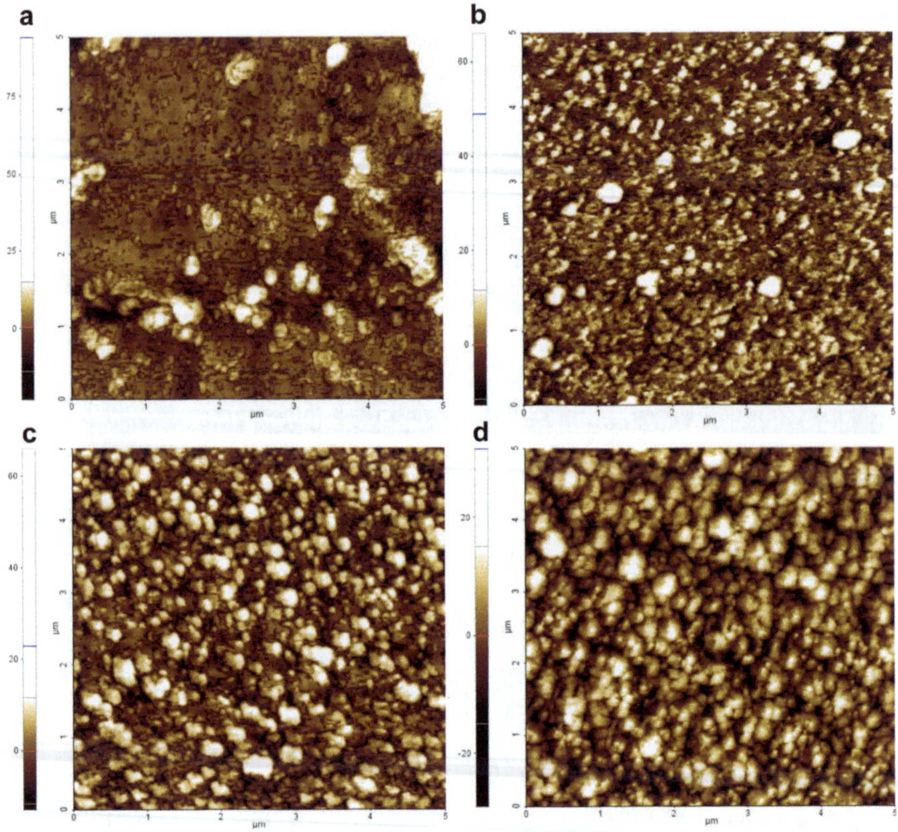

Fig. 5.24 AFM topography of (**a**) bare gold electrode, and gold electrodes after ECV with (**b**) 5 cycles, (**c**) 10 cycles, and (**d**) 30 cycles; $5 \times 5\,\mu m$ scan size, set point 11 nm

5.5 Conclusion

The electrode modified through LBL deposition was unstable and not reproducible. Using SAM and ECV techniques, on the other hand, the nanostructures were firmly immobilized on the gold electrode, resulting in an acceptable stability and reproducibility. CV-electrodeposition of AuNP (ECV) causes an observable increase in the peak current of recorded CV curves compared to bare Au electrodes, causing improved electrode kinetics and a reduction in the oxidation potential (thermodynamically feasible reaction). Moreover, electrodes modified using ECV showed better electron transfer rate compared to other studied methods, making them a better candidate for electrochemical sensing purposes.

Fig. 5.25 AFM height images of the surface of (**a**) bare gold electrode, and gold electrodes after CV electrodeposition with (**b**) 5 cycles, (**c**) 10 cycles, and (**d**) 30 cycles; $5 \times 5\,\mu m$ scan size, set point 11 nm

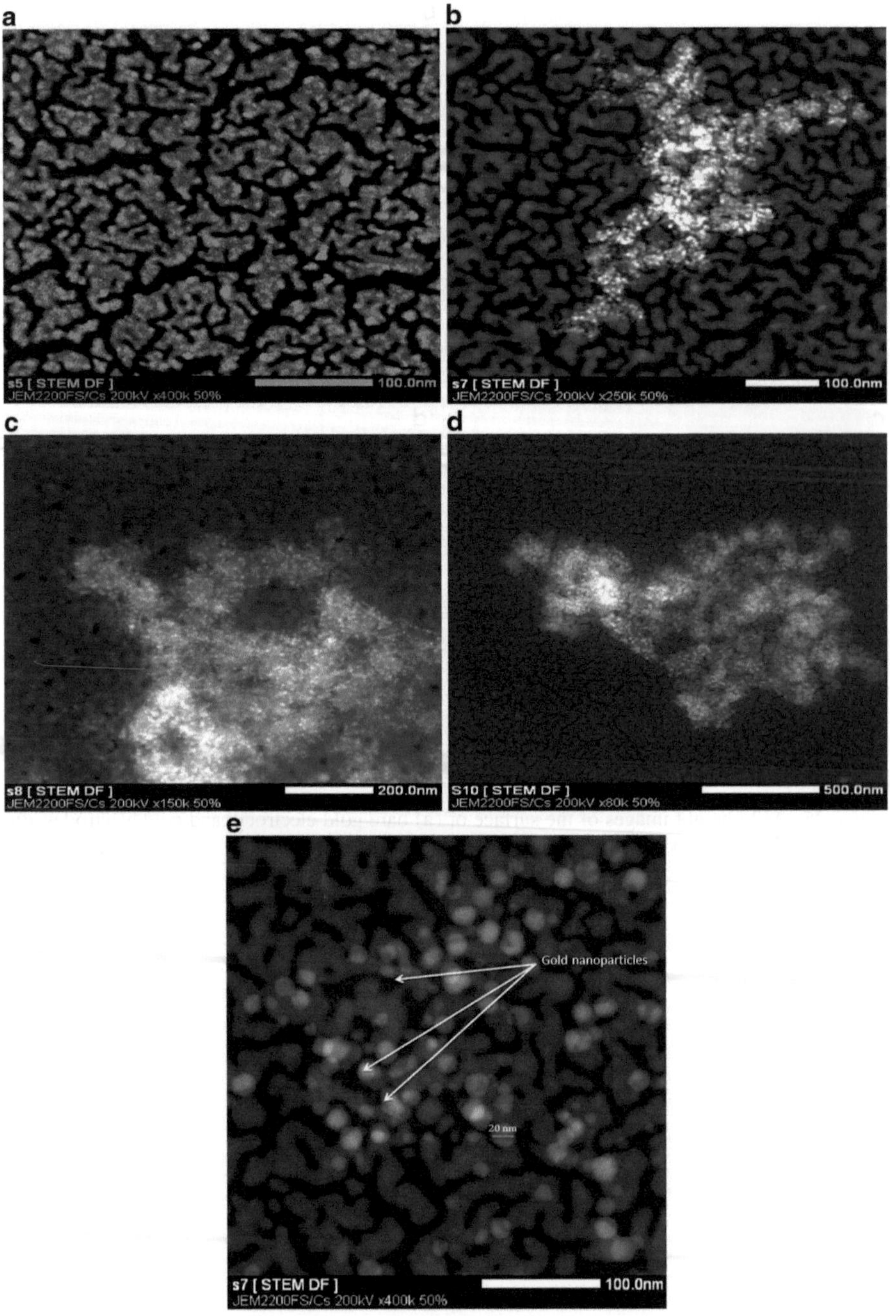

Fig. 5.26 STEM results micrographs of (**a**) bare Au TEM grid before ECV electrodeposition by HAADF contrast, and gold electrodes after ECV electrodeposition with (**b**) 5 cycles, (**c**) 10 cycles, and (**d**) 30 cycles, (**e**) Mean size of the produced nanoparticles by HAADF contrast. (For (**a**), (**b**), (**e**), scale = 100 nm; for (**c**), scale = 200 nm; for (**d**), scale = 500 nm)

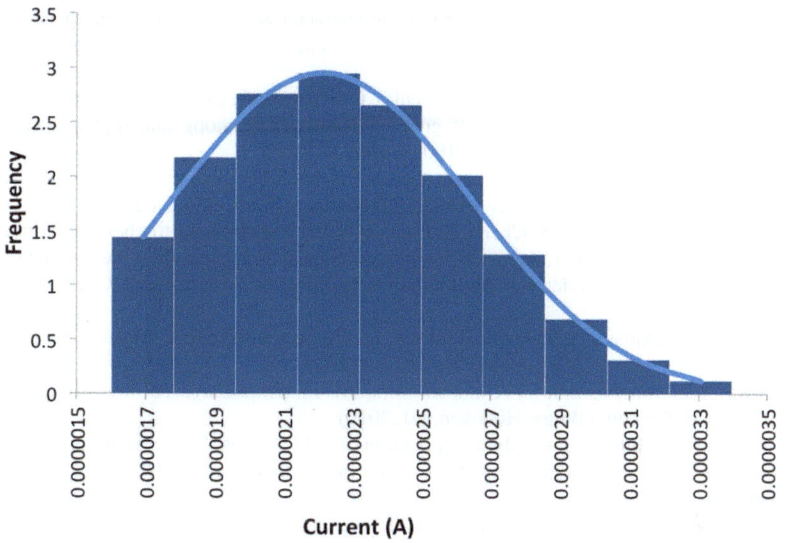

Fig. 5.27 Histogram showing frequency distribution of peak currents measured in a group of electrodes prepared using ECV technique

References

1. M. Ates, A.S. Sarac, Conducting polymer coated carbon surfaces and biosensor applications. Prog. Org. Coat. **66**(4), 337–358 (2009)
2. F. Lucarelli, G. Marrazza, A. Turner, M. Mascini, Carbon and gold electrodes as electrochemical transducers for dna hybridisation sensors. Biosens. Bioelectron. **19**, 515–530 (2004)
3. U.A. Kirgoz, D. Odaci, S. Timur, A. Merkosi, S. Alegret, N. Besun, A. Telefoncu, A biosensor based on graphite epoxy composite electrode for aspartame and ethanol detection. Anal. Chim. Acta **570**(2), 165–169 (2006)
4. R. Sengupta, M. Bhattacharya, S. Bandyopadhyay, A.K. Bhowmick, A review on the mechanical and electrical properties of graphite and modified graphite reinforced polymer composites. Prog. Polym. Sci. **36**(5), 638–670 (2011)
5. R. Navratil, A. Kotzianova, V. Halouzka, T. Opletal, I. Triskova, L. Trnkova, J. Hrbac, Polymer lead pencil graphite as electrode material: voltammetric, XPS and Raman study. J. Electroanal. Chem. **783**, 152–160 (2016)
6. X. Cao, Y. Ye, S. Liu, Gold nanoparticle-based signal amplification for biosensing. Anal. Biochem. **417**, 1–16 (2011)
7. J. Wang, Nanomaterial-based amplified transduction of biomolecular interactions. Small **1**, 1036–1043 (2005)
8. N. Wittenberg, C. Haynes, Using nanoparticles to push the limits of detection. Wiley Interdiscip. Rev. Nanomed. Nanobiotechnol. **1**, 237–254 (2009)
9. K. Nemani, K. Moodie, J. Brennick, A. Su, B. Gimi, In vitro and in vivo evaluation of su-8 biocompatibility. Mater. Sci. Eng. C **33**(7), 4453–4459 (2013)
10. L. Fischer, M. Tenje, A. Heiskanen, N. Masuda, J. Castillo, A. Bentien, J. Emneus, M. Jakobsen, A. Boisen, Gold cleaning methods for electrochemical detection applications. Microelectron. Eng. **86**, 1282–1285 (2009)
11. M. Noh, I. Tothill, Development and characterization of disposable gold electrodes, and their use for lead (ii) analysis. Anal. Bioanal. Chem. **386**, 2095 (2007)

12. Y. Uludag, Z. Olcer, M. Sagiroglu, Design and characterization of a thin-film electrode array with shared reference/counter electrodes for electrochemical detection. Biosens. Bioelectron. **57**, 85–90 (2014)
13. T. Brulle, W. Ju, P. Niedermayr, A. Denisenko, O. Paschos, O. Schneider, U. Stimming, Size-dependent electrocatalytic activity of gold nanoparticles on hopg and highly boron-doped diamond surfaces. Molecules **16**(12), 10059–10077 (2011)
14. M. Finot, G. Braybrook, M. McDermott, Characterization of electrochemically deposited gold nanocrystals on glassy carbon electrodes. J. Electroanal. Chem. **466**(2), 234–241 (1999)
15. H. Penga, Y. Hua, A. Liua, W. Chena, X. Lina, X. Yub, Label-free electrochemical immunosensor based on multi-functional gold nanoparticles–polydopamine–thionine–graphene oxide nanocomposites film for determination of alpha-fetoprotein. J. Electroanal. Chem. **712**, 89–95 (2014)
16. C. Jiang, S. Markutsya, Y. Pikus, V. Tsukruk, Freely suspended nanocomposite membranes as highly sensitive sensors. Nat. Mater. **3**, 721–728 (2004)
17. P. Kohl, *Electrodeposition of Gold, in Modern Electroplating*, ed. by M. Schlesinger, M. Paunovic, 5th edn. (Wiley, Hoboken, NJ, 2010)
18. M. Etesami, N. Mohamed, Catalytic application of gold nanoparticles electrodeposited by fast scan cyclic voltammetry to glycerol electrooxidation in alkaline electrolyte. Int. J. Electrochem. Sci. **6**, 4676–4689 (2011)
19. T. Ling, M. Ahmad, L. Heng, T. Seng, The effect of multilayer gold nanoparticles on the electrochemical response of ammonium ion biosensor based on alanine dehydrogenase enzyme. J. Sens. **2011**, Article ID 754171 (2011)
20. B. van der Zande, M. Böhmer, L. Fokkink, C. Schönenberger, Aqueous gold sols of rod-shaped particles. Phys. Chem. B **101**, 852–854 (1997)
21. T. Green, Gold electrodeposition for microelectronic, optoelectronic and microsystem applications. Gold Bull. **40**(2), 105–114 (2007)
22. M. Mena, P. Sedeno, J. Pingarron, A comparison of different strategies for the construction of amperometric enzyme biosensors using gold nanoparticle-modified electrodes. Anal. Biochem. **336**, 20–27 (2005)
23. M. Backman, N. Juslin, K. Nordlund, Bond order potential for gold. Eur. Phys. J. B **85**, 317 (2012)
24. C. Kaulen, M. Homberger, S. Bourone, N. Babajani, S. Karthäuser, A. Besmehn, U. Simon, Differential adsorption of gold nanoparticles to gold/palladium and platinum surfaces. Langmuir **30**(2), 574–583 (2014)
25. E. Wurster, A. Elbakry, A. Göpferich, M. Breunig, Layer-by-layer assembled gold nanoparticles for the delivery of nucleic acids. Methods Mol. Biol. **948**, 171–182 (2013)
26. Y. Xue, H. Zhao, Z. Wu, X. Li, Y. He, Z. Yuan, The comparison of different gold nanoparticles/graphene nanosheets hybrid nanocomposites in electrochemical performance and the construction of a sensitive uric acid electrochemical sensor with novel hybrid nanocomposites. Biosens. Bioelectron. **29**(1), 102–108 (2011)
27. M.O. Thotiyl, H. Basit, J. Sánchez, C. Goyer, L. Coche-Guerente, P. Dumy, S. Sampath, P. Labbe, J. Moutel, Multilayer assemblies of polyelectrolyte-gold nanoparticles for the electrocatalytic oxidation and detection of arsenic(iii). J. Colloid. Interface Sci. **383**(1), 130–139 (2012)
28. V. Mani, B. Chikkaveeraiah, V. Patel, J. Gutkind, J. Rusling, Ultrasensitive immunosensor for cancer biomarker proteins using gold nanoparticle film electrodes and multienzyme-particle amplification. ACS Nano. **3**(3), 585–594 (2009)
29. H. Chen, Y. Wang, Y. Wang, S. Dong, E. Wang, One-step preparation and characterization of pdda-protected gold nanoparticles. Polymer **47**, 763–766 (2006)
30. S. Basu, S. Ghosh, S. Kundu, S. Panigrahi, S. Praharaj, S. Pande, Biomolecule induced nanoparticle aggregation: effect of particle size on interparticle coupling. J. Colloid. Interface Sci. **313**, 724–734 (2007)

31. H. Chen, G. Zhao, Construction of ionic liquid self-assembled monolayer on gold electrode for hemoglobin modifying and peroxide biosensing. Int. J. Electrochem. Sci. **7**, 12907–12921 (2012)
32. T. Luczak, Gold and nanogold electrodes modified with gold nanoparticles and meso-2,3-dimercaptosuccinic acid for the simultaneous, sensitive and selective determination of dopamine and its biogenic interferents. Electroanalysis **26**(10), 2152–2160 (2014)
33. T. Luckaz, Comparison of electrochemical oxidation of epinephrine in the presence of interfering ascorbic and uric acids on gold electrodes modified with s-functionalized compounds and gold nanoparticles. Electrochim. Acta **54**, 5863–5870 (2009)
34. C. Vericat, M. Vela, G. Benitez, P. Carrob, R. Salvarezza, Self-assembled monolayers of thiols and dithiols on gold: new challenges for a well-known system. Chem. Soc. Rev. **39**, 1805–1834 (2010)
35. M.D. Millone, H. Hamoudi, L. Rodriguez, A. Rubert, G.A. Benítez, M.E. Vela, R.C. Salvarezza, J.E. Gayone, E.A. Sanchez, O. Grizzi, C. Dablemont, V.A. Esaulov, Self-assembly of alkanedithiols on au(111) from solution: effect of chain length and self-assembly conditions. Langmuir **25**, 12945–12953 (2009)
36. J. Tour, L. Jones II, D. Pearson, J.J.S. Lamba, T.P. Burgin, G. Whitesides, D.L. Allara, A.N. Parikh, S.V. Atre, Self-assembled monolayers and multilayers of conjugated thiols, α, ω-dithiols and thioacetyl-containing adsorbates. understanding attachments between potential molecular wires and gold surfaces. J. Am. Chem. Soc. **117**, 9529 (1996)
37. H. Ron, S. Matlis, I. Rubinstein, Self-assembled monolayers on oxidized metals. 2. Gold surface oxidative pretreatment, monolayer properties, and depression formation. Langmuir **14**, 1116–1121 (1998)
38. Y. Tang, P.G. Chen, Nanoparticle-electrodeposited electrodes used for p-nitrophenol detection in acidic media: effect of electrodeposition parameters on particle density, size distribution, and electrode performance. J. Chin. Chem. Soc. **58**(6), 723–731 (2011)
39. Y. Takahashi, H. Umino, S. Taura, S. Yamada, An electrochemical approach for fabricating organic thin film photoelectrodes consisting of gold nanoparticles and polythiophene. Rapid Commun. Photosci. **2**(3), 79–81 (2013)
40. J. Wu, L. Li, B. Shen, G. Cheng, P. He, Y. Fang, Polythymine oligonucleotide-modified gold electrode for voltammetric determination of mercury (ii) in aqueous solution. Electroanalysis **22**, 479–482 (2010)
41. B. Park, D. Kim, D. Yoon, Surface modification of gold electrode with gold nanoparticles and mixed self-assembled monolayers for enzyme biosensors. Kor. J. Chem. Eng. **28**(1), 64–70 (2011)
42. E. AlShamaileh, H. Saadeh, V. Favry, Modification of gold surface with gold nanoparticles and cyclohexyl dithiocarbamate as a selective sensor for cysteine. J. Chem. **2013**, Article ID 383921 (2013)
43. N. German, A. Ramanavicius, A. Ramanaviciene, Electrochemical deposition of gold nanoparticles on graphite rod for glucose biosensing. Sens. Actuators B **203**, 25–34 (2014)
44. K. Kerman, M. Vestergaard, E. Tamiya, Electrochemical DNA biosensors: protocols for intercalator-based detection of hybridization in solution and at the surface, in *Methods in Molecular Biology: Biosensors and Biodetection*, vol. 504 (Humana Press, New York, 2009)
45. P. Khashayar, G. Amoabediny, B. Larijani, M. Hosseini, R. Verplancke, D. Schaubroeck, M. De Keersmaecker, M. Adriaens, J. Vanfleteren, Characterization of gold nanoparticle layer deposited on gold electrode by various techniques for improved sensing abilities. Biointerface Res. Appl. Chem. **6**(4), 1380–1390 (2016)

Chapter 6
Oc and CTX BioSensors: Characteristics and Validation

Abstract As mentioned in Chap. 3, while immunoassays have an extensive application in both clinic and research, the immunosensor concept is yet to be implemented in routine diagnostics [23, 24]. There are certain issues in this regard, the most important of which is the coupling chemistry of proteins on surfaces, which is critical for optimal functioning. In other words, the optimal orientation of Ab molecules on the surface is of great importance in selective and sensitive detection in a reproducible way without affecting its activity [25]. The relative lack of long-term stability of biological molecules is the most important limitation in commercializing biosensors.

Therefore, this chapter deals with the immobilization of two main BTMs (Oc and CTX) on the glass-based gold electrode fabricated as mentioned in Chap. 5 and coated with AuNPs. To our knowledge, a biosensor to assess serum levels of Oc has not been developed before. A stepwise protocol to develop these sensors is explained here.

6.1 Introduction

Immobilization methods vary largely based on the surface and protein properties. Ideally, as mentioned earlier, the protein conformation should remain intact during the process so that the protein's function would be retained. Moreover, the active sites of Ab should be accessible to reaction partners in order to have a high-performance, reproducible assay [26]. This is while once immobilized, biomolecules adsorbed on AuNPs are not free and adopt certain orientation and conformation on the surface [28]. This raises the question of how the activity of an antibody is affected by conjugation with a gold particle. Although protein-conjugated gold nanoparticles have been successfully employed as biological markers for decades, instances of total or partial loss of specific activity of a protein upon conjugation have also been reported [27].

This chapter contains information published in the article from the author and her team in "Sensors" [29].

P. Khashayar, *Point-of-Care Solution for Osteoporosis Management*,
DOI 10.1007/978-3-319-55053-4_6

6.2 Oc Biosensor

6.2.1 Oc

Oc, also known as bone gamma-carboxyglutamic acid-containing protein (BGLAP), is the most abundant non-collagenous protein in bone, comprising 1–2% of the total protein content [1]. It is a small 44-amino acid long protein (5.8 kDa) synthesized exclusively by osteoblasts (and odontoblasts). The specific glutamyl residues at positions 17, 21, and 24 have a second carboxylate group that forms γ-carboxyglutamyl residues (Gla), which stabilize the α-helical portion. When Oc meets calcium ions, these active sites (3 Gla residue and 1 Asp) chelate to the ions and Oc folds to a special structure.

Oc is released into the circulation from the matrix during bone resorption and, therefore, its high levels are considered as a marker of bone turnover [2]. Previous studies suggest that serum Oc levels, even in the absence of alkaline phosphatase, is useful in monitoring changes that currently cannot be assessed with BMD assessment in osteoporosis treatment [3, 4]. As mentioned in Chap. 2, Oc is a specific, sensitive, and promising marker to be used for better prognosis of osteoporosis and for monitoring response to antiresorptive therapy [5] (Fig. 6.1).

6.2.2 Sensor Fabrication Process

As mentioned earlier, the glass-based gold electrodes fabricated based on the method explained in Chap. 5 were used. In order to have an AuNP layer on the electrodes that would allow Ab molecules to immobilize on the surface, the ECV technique was used.

The final goal of immobilization strategies is to improve assay sensitivity, specificity, reproducibility, and even application. In this regard, different techniques were modified to form stable and strong bonds between the protein and the immobilization surface, minimize non-specific adsorption of biomacromolecules, keep the proteins in an active state, and finally orient proteins for unhindered access of binding partners. Consequently, controlled experiments against the standards were performed for verification of the best method and conditions of immobilization of Oc antibody on the gold electrodes. The applied protocols are shown in the flowchart in Fig. 6.2. The first two protocols were based on physisorption immobilization, whereas the others were based on covalent attraction between constituents using different cross-linkers.

In the first protocol, fragment antigen-binding (Fab) antibody fragments were deposited onto gold via native thiol groups. Passive absorption does not provide permanent attachment of the coating as molecules may desorb from the surface over time. Proteins may also lose their properties after being absorbed into the surface, caused by changes in tertiary structure or binding of active sites to the surface

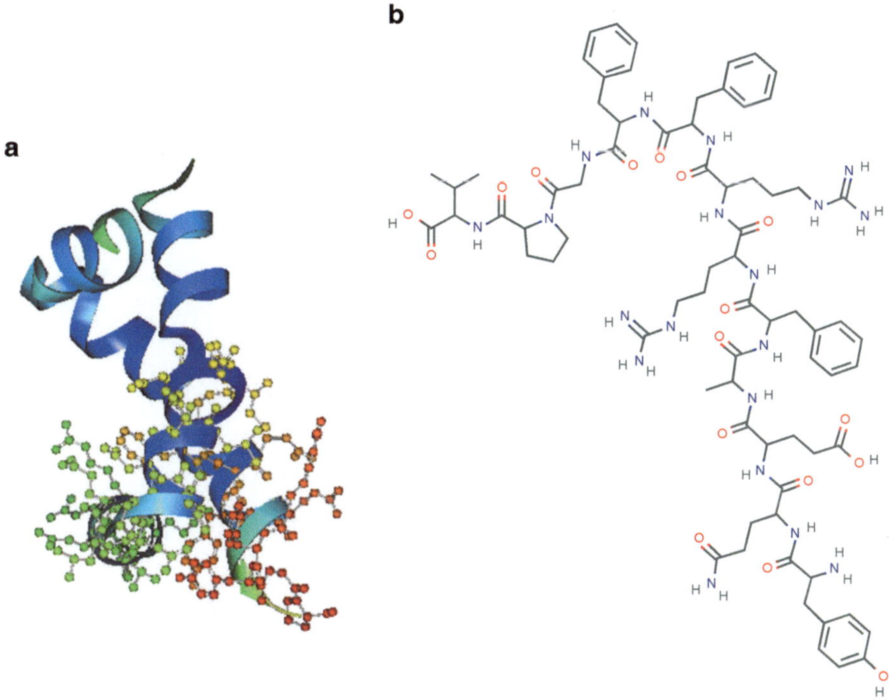

Fig. 6.1 3D and 2D structure of Oc. "From National Center for Biotechnology Information. PubChem Compound Database; CID=102601936, https://pubchem.ncbi.nlm.nih.gov/compound/102601936 (accessed Jan. 5, 2017)"

rendering it inaccessible. In the second protocol, still the bonding was not strong enough and thus the antibody was flushed away during the washing steps.

As for other protocols, covalent coupling using different combinations of EDC and sulfo-NHS was applied. These reactions are not only specific and controllable but also the number of covalent bonds between the surface and Ab can be optimized to help maintain the tertiary structure of the protein.

Here we explain the most complex process (protocol 3), with the maximum number of steps. The other protocols were tested similarly using the same concentrations and incubation times mentioned in this protocol, skipping the non-mentioned materials.

In protocol 3, appropriate amount of a 10 mM solution of GSH was deposited on the gold surface and left to dry at room temperature for an hour [6]. GSH bonds with gold through the formation of an Au-S bond, while providing carboxylate groups for the next layers to react with. A mixture of EDC/sulfo-NHS was used to facilitate the bonding between the carboxylic acid groups of the monolayer and the amine groups of Ab. In this regard, the freshly prepared EDC and sulfo-NHS were mixed with a volume ratio of 1:1 and incubated for 20 min. Appropriate amount of Ab was then

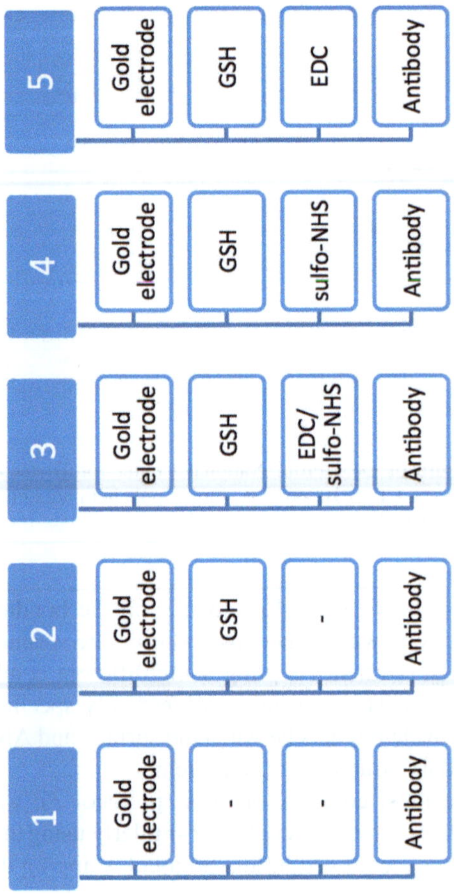

Fig. 6.2 Flowchart depicting experimental design for immobilizing antibody on the electrode

added to the solution and incubated for another 60 min. After gently rinsing the prepared surface with distilled water, the electrodes were covered with the solution of EDC/sulfo-NHS/Ab and incubated for 180 min at room temperature.

In order to minimize non-specific adsorption of off-target biomolecules and remove non-adsorbed proteins, the surface was gently rinsed. Effective removal of excess EDC/sulfo-NHS after activation is important for preventing protein cross-linking.

Then with the aim of blocking the uncoated surface, the unreacted active functional groups were blocked or deactivated by BSA. 20 μl of 0.2% BSA solution in PBS buffer was added for 1 h. The electrode was then rinsed for 3 min with 0.1 M PBS (coupling buffer) and then 0.1 M PBS-Tween 20 (washing buffer). PBS-Tween 20 is commonly used to minimize the background noise from non-specific binding in bioassays. The electrodes were then shaken to remove excess water.

While DI water and PBS, used as the rinsing reagents in our experiment, are inert, care was taken while applying these solutions to maintain the integrity of the electrode surface as the detachment of the conjugates can be an issue.

The Ab-Ag reaction was then tested by the introduction of different concentrations of Ag solution using a fine pipette.

6.2.3 Characteristics of the Oc Electrochemical Chip

6.2.3.1 Surface Modification

Successful electrodeposition of gold nanoparticles on the surface and immobilization of antibody on the modified surface was verified by STEM (Fig. 6.3).

To further analyze the chemical composition of different layers, XPS characterization was employed at different steps. In the first step, the XPS signature of the S2p doublet (1639–1651 eV) was suggestive of the attachment of GSH to the gold surface through the thiol group. The N1s and C1s core level spectrums noted in the successive steps mainly originated from the antibody. The presence of satellite (293 eV) and aromatic (284.3 eV) lines in the XPS spectrum also confirmed the immobilization of the antibody on the surface has been successful.

6.2.3.2 Electrochemical Impedance Spectroscopy

As mentioned in Chap. 3, EIS is another powerful method, commonly used in electrochemical research, to analyze the electrical resistance of a system. The technique is sensitive to surface phenomena and changes of bulk properties. Therefore it was applied within this work to characterize surface modifications and detect binding events on the transducer surface. The assembly of the layers on the gold electrode surface and the loading of AuNP were confirmed by EIS measurements.

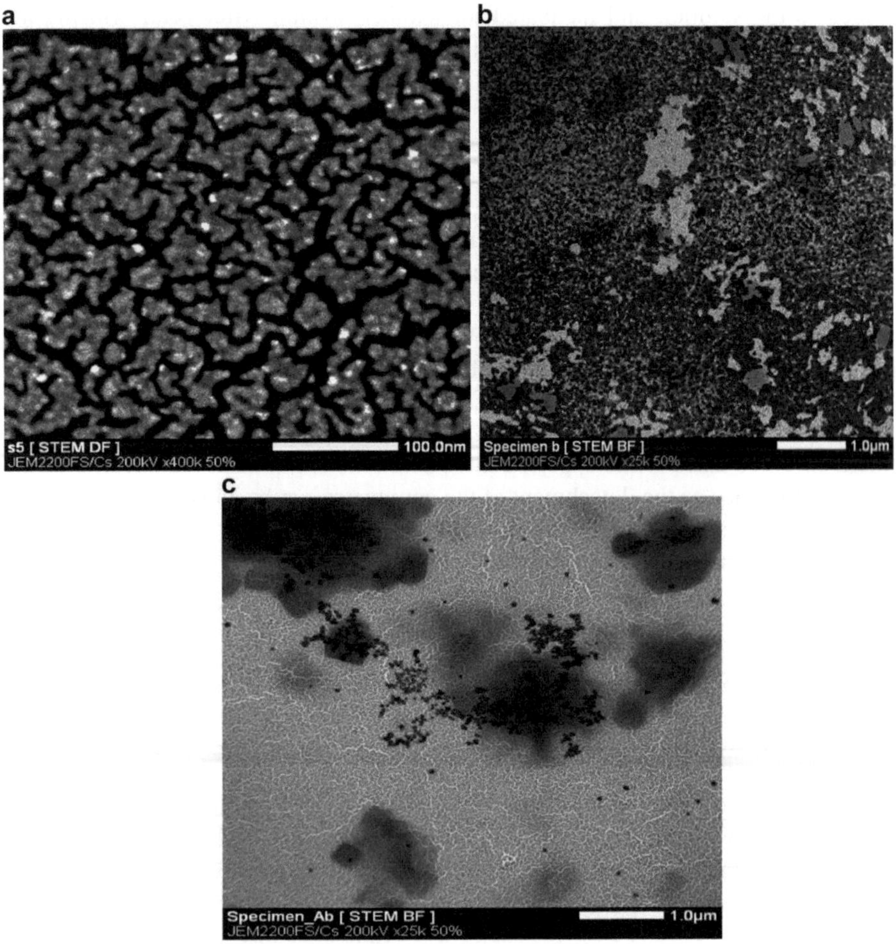

Fig. 6.3 STEM micrographs of (**a**) bare gold TEM grids by HAADF contrast, (**b**) AuNP-modified gold TEM grids by BF contrast, (**c**) Ab-functionalized gold TEM grids by BF contrast. (For (**a**), scale = 100 nm; for (**b**), scale = 200 nm; for (**c**), scale = 1.0 μm)

Figure 6.4 shows a Nyquist plot of impedance for the stepwise modification process of the gold electrode. The data was obtained by using 0.1 mM $K_3[Fe(CN)_6]$ and upon application of the biasing potential 0.17 V and 5 mV amplitude in the frequency range of 0.1–100 kHz. As already known, the impedance spectra contain a semicircle portion at higher frequencies relating to the electron transfer limited process and a linear part at lower frequencies corresponding to diffusion. The diameter of the semicircles showed the charge transfer resistance rate.

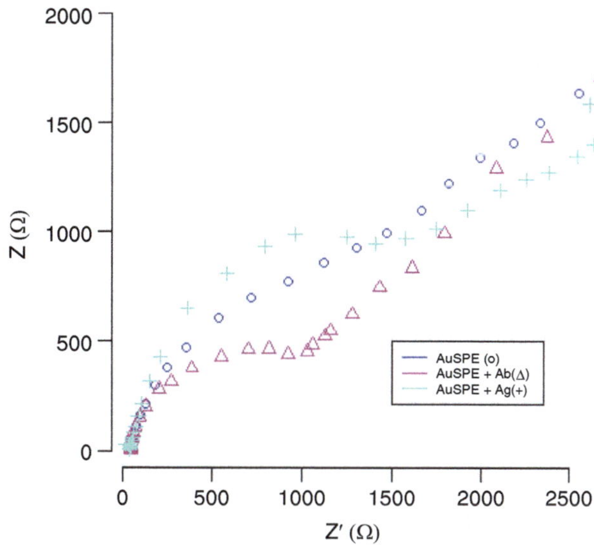

Fig. 6.4 EIS results for each modification step (*Marine* = gold electrode with gold nanoparticles capped with GSH, *Pink* = gold electrode with antibody conjugated gold nanoparticles, *Blue* = sensor + Ag)

In the EIS measurement, the negatively charged ferri/ferrocyanide couple $[Fe(CN)_6]^{3-/4-}$ was employed as the redox marker for electrochemical characterization, and thus the diffusion of $[Fe(CN)_6]^{3-/4-}$ and mass loading of the antigen onto the surface of the electrodes have a significant effect on the charge transfer resistance.

As can be seen from Fig. 6.4, the bare electrode produces an almost straight line in the Nyquist plot with no semicircle domain, implying a diffusion-limited electrochemical process of the redox coupling at the electrode surface. After adding additional layers, a semicircle was noted, indicating the increase of the interfacial charge transfer resistance. Considering the nature of the layers, the changes in this diameter could be explained. Ab binding caused a physical barrier, limiting the access of the redox probe to the electrode. Exposure of Ag generates a further barrier for the electrochemical process, causing a slight increase in the transfer resistance.

6.2.4 Analytical Procedure

Ag concentration against current read during DPV analysis in each protocol was used to calculate and plot linear and logarithmic fit R-squares (degree of fit). The DPV measurements in this work were performed in 0.1 mM $K_3[Fe(CN)_6]$, containing 0.01 M NaCl, applying following parameters: 0.025 V modulation amplitude, 0.05 s modulation time, 0.005 V step potential, and voltage range from 0.5 to 1.1 V. The results showed that the immobilization was most effective for sulfo-NHS activation (protocol 4) as it resulted in not only a decreasing trend suggestive of the inverse correlation of current and Ag concentration, but also the best R-square.

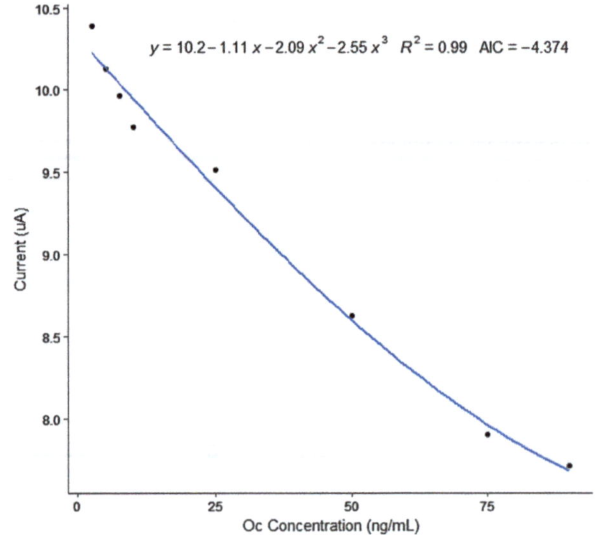

Fig. 6.5 Sensor responses to varying Ag concentrations (2.5–90 ng/mL) using protocol 4

The stability of the immobilization process was also studied by washing the electrodes with PBS for four times. Again electrodes prepared using protocol 4 showed the best results. In other words, the results showed that the immobilization was most successful when sulfo-NHS activation was applied and that the best results were achieved when carboxylates (-COOH), found on the surface after the deposition of GSH on the gold electrode, reacted with sulfo-NHS (Fig. 4.5). The resulted NHS ester then reacted with primary amines (-NH2) of the antibody to form stable amide cross-links.

Afterwards, through plotting the current in that protocol against different concentrations, the calibration curve was formed. As mentioned earlier, as more antigen was introduced onto the surface upon reaction with the immobilized antibody, the redox marker diffusion to the surface of the electrode was hindered due to repulsive interactions. Thus lower current was observed.

Figure 6.5 illustrates the calibration curve plotted using average response of the sensor towards its target, Oc Ag, from experiments performed under the conditions described above.

The calibration curve was generated based on the average value of measurements conducted in triplet sets on the sensing interface, which was treated with different amounts of antigen from the lowest to the highest concentrations. Parameters such as lower limits of detection (LLD), upper limits of detection (ULD), and consequently a dynamic range were calculated. Other assessed parameters included limit of detection (LoD) and limit of quantification (LoQ), sensitivity (slope of the gradient), reproducibility (standard deviation divided by the average in percent), tightness of fit (R-square of trend line applied), and behavior (linear, non-linear, etc.).

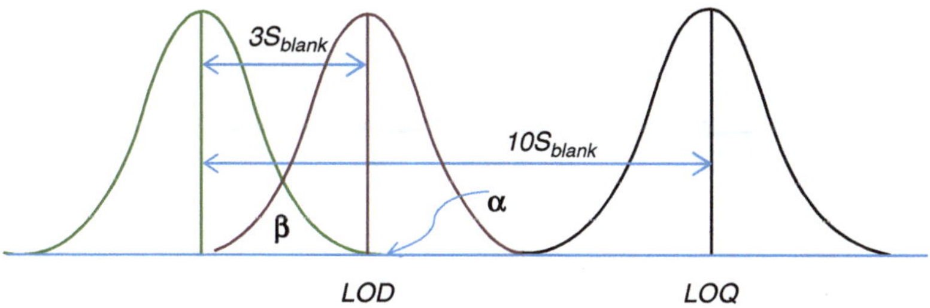

Fig. 6.6 Illustration of the concept of LoD and LoQ by showing the theoretical normal distributions associated with blank, LoD, and LoQ level samples

The LoD and LoQ were calculated using the following equations (Eqs. (6.1), (6.2)):

$$\text{LoD} = \frac{3.3 * \delta}{S} \tag{6.1}$$

$$\text{LoQ} = \frac{10 * \delta}{S} \tag{6.2}$$

where δ is the standard deviation of the response and S is the slope of the calibration curve [7].

Figure 6.6 illustrates the relationship between the blank, LoD, and LoQ. For a signal at the LoD, the alpha error (probability of false positive) is small (1%) [8]. However, the beta error (probability of a false negative) is 50% for a sample that has a concentration at the LoD, suggesting that the sample could contain an impurity at the LoD, but there is a 50% chance that the measurement would give a result less than the LoD. At the LoQ, there is minimal chance of a false negative.

The magnitude of electrochemical response current was linearly dependent on Oc concentration with an R^2 of 0.99 over the concentration range. The detection limit was calculated to be 0.65 ng/mL. The device had an LoQ of 1.97 ng/mL with a correlation coefficient of 0.99. The sensor was capable of assessing Ag levels in the range of 2.5–90 ng/mL.

A schematic illustration of the preparation process of the immunosensor is shown in Fig. 6.7.

6.2.5 Real Serum Measurement

Concentration gradients of serum Oc levels were tested using ECLIA and ran against the number read by the sensor for verification. In this regard, the serum of eight patients whose Oc levels were recently measured using ECLIA was used. Based on the plotted calibration curve and using the trend function in Excel, the concentration of Oc in the tested samples was calculated. The results were then compared to that of ECLIA.

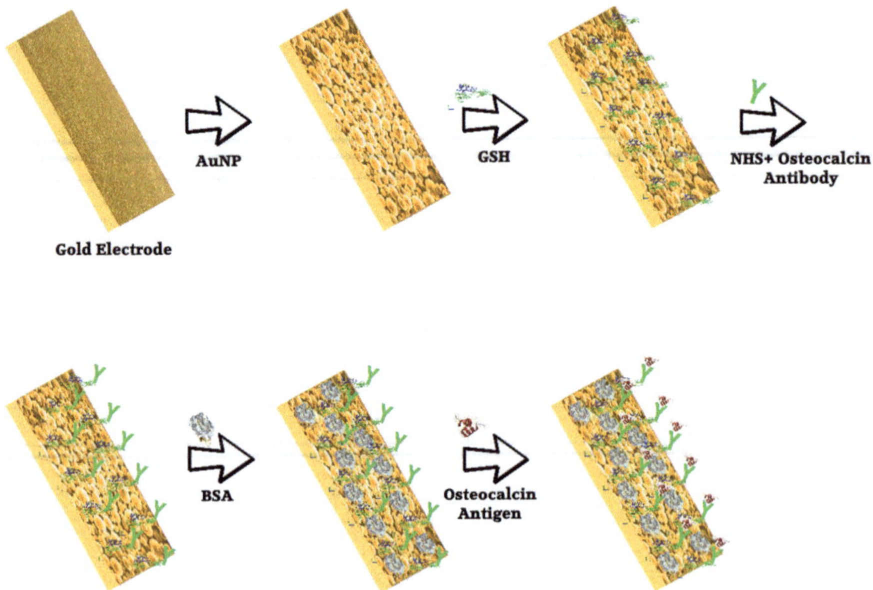

Fig. 6.7 A Schematic illustration of the preparation process of the Oc immunosensor

Table 6.1 Precision of Oc assay using ECLIA and Oc-sensor

Serum samples	ECLIA results (ng/mL)	Oc-sensor results (ng/mL)
1	10.3	10.33
2	15	14.71
3	17.1	17.90
4	26.8	24.35
5	29.8	26.88
6	32.1	29.35
7	33.7	32.11
8	45.5	46.16

According to the results presented in Table 6.1, the coefficients of variation for the ECLIA and Oc sensor were calculated to be 4.4% and 4.48%, respectively.

The correlation between our sensor's results and that of ECLIA is shown in the correlation and Bland and Altman plots in Fig. 6.8. The equation of the trend line in the correlation plot has a slope of 0.97 and R^2 of 0.98. The mean difference between the two methods according to the Bland and Altman plot was 1.06 ng/mL and the 2s limits of the standard deviation (2SD) was 3.03 ng/mL. This indicates that the Oc sensor can measure Oc levels in serum sensitively and specifically [9].

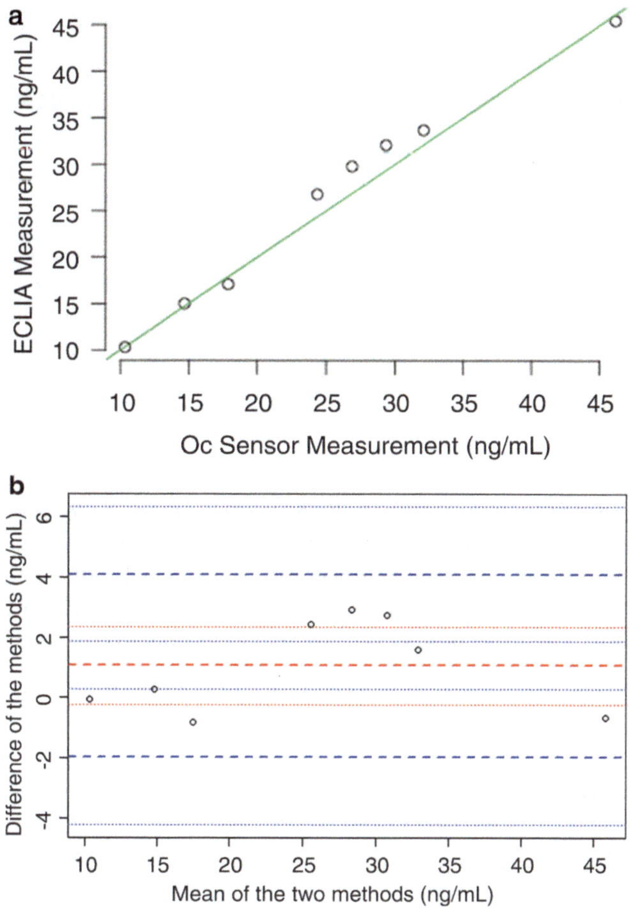

Fig. 6.8 Statistical analysis comparing ECLIA vs. Oc sensor immunoassay. (**a**) Correlation plot, (**b**) Bland and Altman plot

6.2.6 *PDMS-Stamps*

Previous studies have considered microcontact printing as an effective technique for parallel transfer of material to another surface through stamping thousands of microspots of the same molecule simultaneously over areas larger than 1 cm or developing complex protein pattern possible [10, 11]. In order to improve the functionalization process, a combination of hydrophobic and covalent interactions were applied (Fig. 6.9). As described earlier, the simplicity of physisorption makes it a preferred method for immobilizing proteins in early, proof-of-concept experiments. In this regard, PDMS stamps were used.

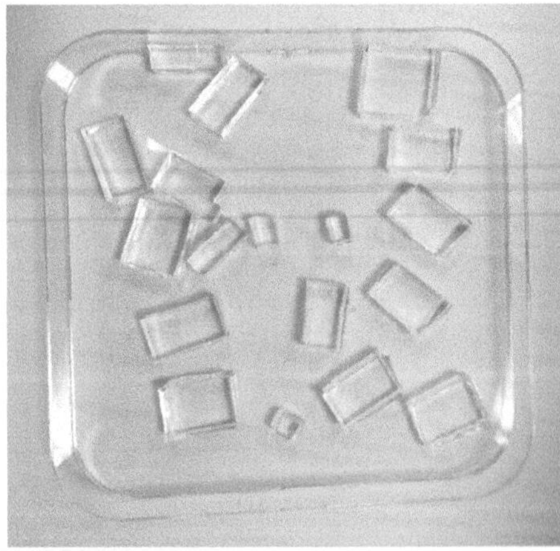

Fig. 6.9 PDMS-stamps used for microcontact printing of the antibodies on the electrode surface

The efficacy of the technique, which benefits from an entropically driven interaction and hydrophobic adsorption rather than the formation of chemical bonds, depends on well-known experimental variables such as pH, salt concentration, and temperature. It also requires multiple inking and stamping steps in order to develop discrete pattern of multiple protein species, and according to our experience was not reproducible.

6.3 CTX Biosensor

6.3.1 CTX

Type I collagen accounts for 90% of the organic matrix of the bone, and is subjected to a series of enzymatic and nonenzymatic intra- and extracellular post-translational modifications that may induce bone strength during the normal aging process or on the pace of a disease [12]. Collagen cross-linking is, therefore, of great importance in determining the biomechanical competence of the bone [13]. Several studies have related the bone content of C-telopeptide (CTX) cross-links, in part independent of the bone mineral value, with bone strength, suggesting serum CTX as a reference bone resorption marker [14, 15]. In other words, despite the marked effects of circadian variation and other factors such as diet on serum levels of CTX, the marker is reported to be sensitive to short-term changes in bone metabolism.

The first serum CTX assay was a competitive polyclonal antibody ELISA [16, 17]. To overcome the limitations linked to ELISA, several label-free immunosensing biosensors for the detection of CTX levels were developed but never commercialized [18, 19].

6.3.2 *Sensor Fabrication Process*

Again, the gold electrodes prepared in Chap. 5 and coated with an AuNP layer using ECV technique were used. The immunosensor was fabricated by the immobilization of antibodies on the gold electrode through EDC/sulfo-NHS covalent-linking.

Firstly, 4 μL of a 10 mM GSH solution was coated on the working electrode and dried at room temperature for 1 h. After gently rinsing with distilled water, the gold electrodes functionalized with carboxylic groups were incubated in a mixture solution of 400 mM EDC and 100 mM sulfo-NHS and 10 μg/mL CTX for 2.5 h. In order to make this solution, EDC and sulfo-NHS (1:1 ratio) were first reacted with each other for 20 min and then CTX Ab was added to the solution and incubated for another 60 min. Subsequently, the electrodes were rinsed to remove the excess antibody complex.

In order to avoid non-specific binding, 10 μL of BSA solution (0.2% m/v in 0.1 M PBS, pH 7.4) was deposited on each electrode for 1 h. Finally, the electrodes were rinsed with 0.1 M PBS and 0.1 M PBS-Tween 20. The electrodes were then shaken to remove excess water. Modified electrodes were stored dry until further use with actual antigen samples, but not longer than one day. The electrodes carrying the conjugated antibodies were treated carefully as the detachment of the conjugates can be troublesome if proper handling/pipetting is not employed during washing/blocking steps.

6.3.3 *Characteristics of CTX Electrochemical Chip*

6.3.3.1 Surface Morphology

We used STEM to determine three-dimensional surface morphology of the electrodes after every modification step and to study the interaction between CTX antibody and the electrode surface. STEM images of the fabricated Au nanoparticles on gold surface demonstrated highly uniform distribution of AuNP of approximately 20 nm in diameter on the surface following electrodeposition (Figs. 6.10, 6.11, and 6.12). As a result, total surface roughness, and thus surface area of the electrode was enhanced. The observation of Ab-functionalized electrodes indicated the affinity of AuNPs to CTX antibody, and the aggregation properties of the antibody complex.

XPS similarly confirmed the successful modification of the electrode surface in every step.

6.3.3.2 Electrochemical Behavior

The AuNP layer enhanced the electrochemical conductivity of the AuNP-modified electrode compared to the bare one, which is related to the increased active surface

Fig. 6.10 (**a**) STEM micrographs and (**b**) CV of a bare gold TEM grid

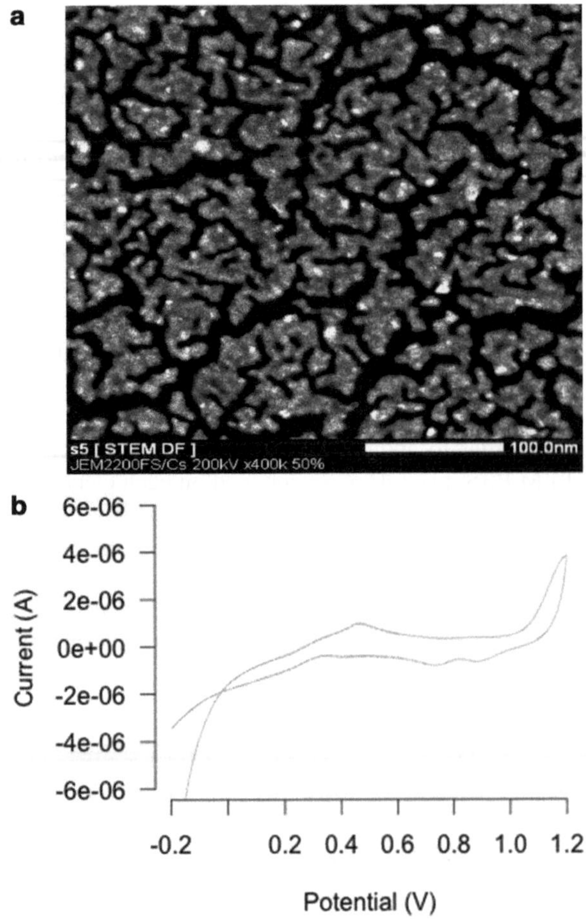

area of the electrode (Figs. 6.10, 6.11, and 6.12). As mentioned in Chap. 5, the real electrochemical surface area of the gold electrode was 2.3 times greater than the area before electrodeposition, suggesting that the technique resulted in a bigger surface for antibody immobilization.

In addition, the CV of AuNP-modified electrodes showed a remarkable oxidation peak at 0.142 V and a reduction peak at 0.9 V. This sharp oxidation peak is due to oxide formation and the occurrence of Au stripping. As a result, the AuNP-modified electrode displayed an advantage for signal amplification.

After the addition of the antibody complex, however, a significant signal reduction was noted in the electrical response. This is because of the binding of antibody on the electrode surface, which forms a hydrophobic surface that retards the interfacial electron transfer, a condition that will be reflected as reduced electrical response.

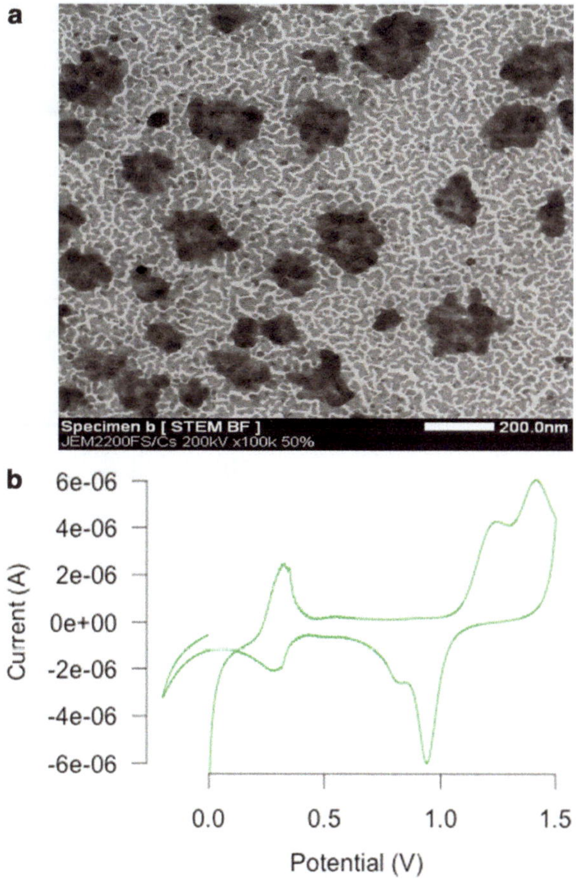

Fig. 6.11 (**a**) STEM micrographs and (**b**) CV of an AuNP-functionalized gold TEM grid (External reference and auxiliary electrodes were used for assessing AuNP-modified electrodes)

6.3.4 Analytical Procedure

The fundamental structural unit of an antibody is two identical heavy and light chains interconnected by disulfide bridges. The chains are entirely composed of constant and variable regions, combining one interaction site for the antigen. The conformation of amino acids of antibody in this site determines its antigen binding activity [20]. Several non-covalent bonding, such as van der Waals and non-polar hydrophobic interactions as well as London dispersion attractive and steric repulsion forces, contribute to the molecular interactions between the antibody and antigen.

The antibody–antigen reaction on our electrodes was tested by the introduction of Ag solution using a fine pipette. In this regard, concentration gradients of the target, CTX, were prepared and ran against the sensor for verification. The concentration against current read during DPV analysis was used to calculate and plot linear and logarithmic fit R-squares (degree of fit).

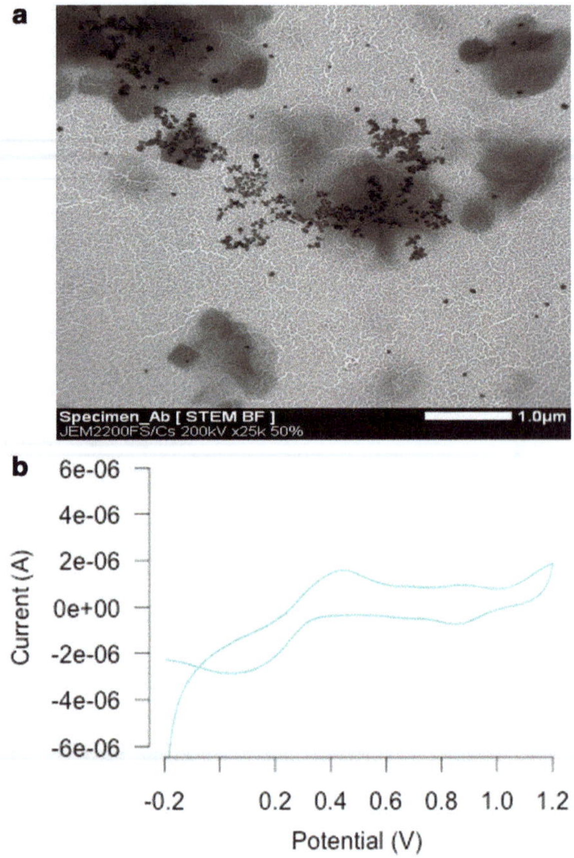

Fig. 6.12 (**a**) STEM micrographs and (**b**) CV of an Ab-functionalized gold TEM grid

Under the optimum conditions, the DPV peak current of the immunosensor array for detection of CTX decreased proportionally with the increasing concentration of analytes (Fig. 6.13). A linear response was observed over the logarithm of concentration range (1–1000 pg/mL) tested and the activity of the receptor-antigen couple was verified. A slope of 0.004 and a linear regression constant (R^2) of 0.96 were obtained from the response curve. The correlation coefficient was 0.82 and the LOQ was calculated to be 16.5. The LoD was found to be 5.44 pg/mL, which was lower than the normal CTX range in plasma (50–450 pg/mL), indicating the sensitivity of the assay.

A schematic illustration of the preparation process of the immunosensor is shown in Fig. 6.14.

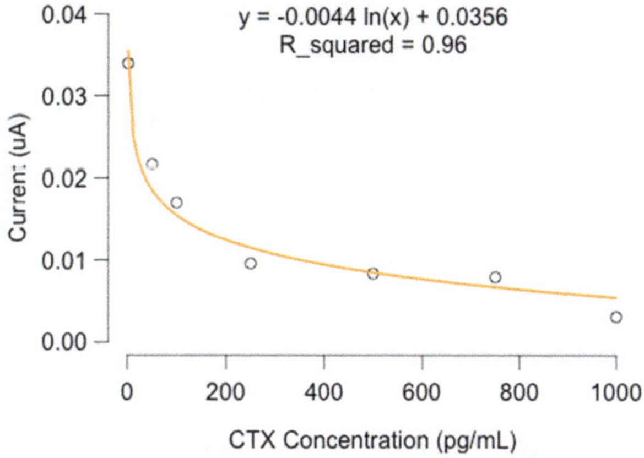

Fig. 6.13 Calibration curve of different CTX concentrations using the proposed immunosensor for signal tracing and amplification

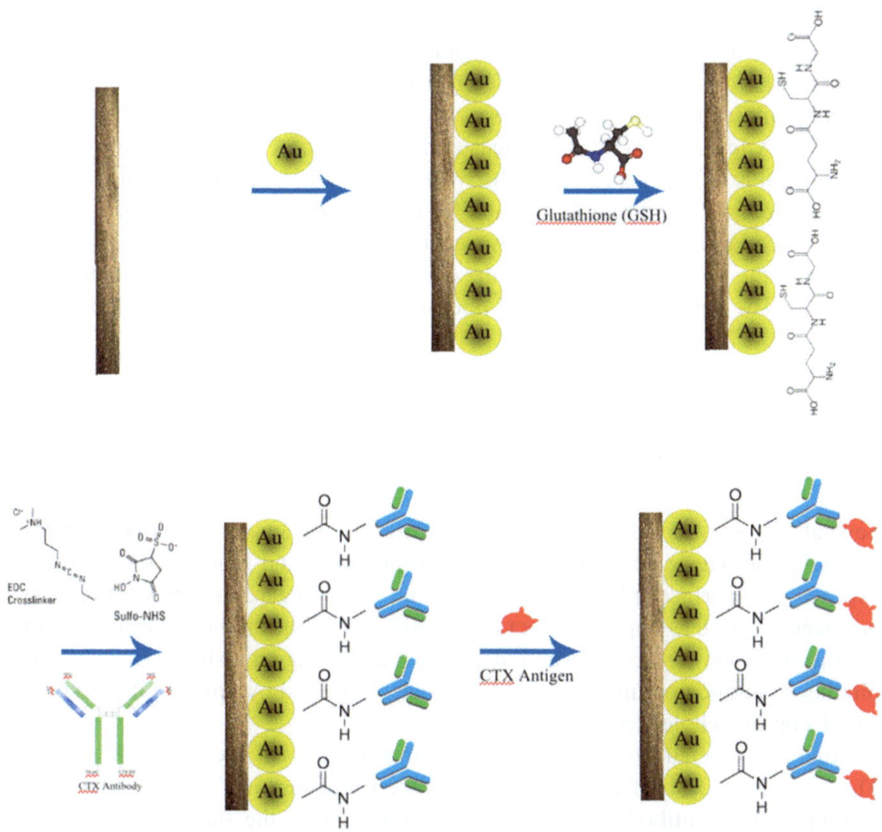

Fig. 6.14 Surface preparation is typically made possible by modifying the gold surface by AuNP, leaving carboxyl groups on the surface which in turn can be activated using zero crosslength linkers such as (EDC/NHS), rapidly accepting an antibody and detection can occur under target binding

Table 6.2 Assay results of clinical serum samples using the proposed and reference (ECLIA) methods

Serum samples	Proposed method (pg/mL)	Reference method (pg/mL)	Relative error (%)
1	268.02	290	7.58
2	403.13	430	6.25
3	588.43	570	3.23
4	216.38	230	5.92
5	451.03	420	7.39
6	549.55	520	5.68
7	766.34	810	5.39

6.3.5 Real Serum Measurement

To evaluate the reliability and application potential of the proposed immunoassay method, the assay results of CTX in human serum samples using the proposed method were compared with the reference values obtained by commercial electro-chemiluminescent single-analyte tests. In this regard, the serum of seven patients whose CTX levels were recently measured using ECLIA was used. The results are listed in Table 6.2. Acceptable results with relative errors less than 7.58% indicated good accuracy of the proposed method for real sample analysis.

The correlation between our sensor results and that of ECLIA was also investigated and the results are shown in the correlation and Bland and Altman plots in Fig. 6.15. The equation of the trend line in the correlation plot has a slope of 0.98 and R^2 of 0.97.

6.4 Optimization

Several factors affect the LoD and thus sensitivity of an assay. Reducing the risk of loss of activity of the antibody upon conjugation and non-specific binding on exposed gold surfaces plays an important role in this regard. In this immunosensor, however, we tried to improve direct immobilization of the antibody through modifying the electrode surface by AuNP deposition and using cross-linkers. Non-specific binding of antigen on uncovered gold was also resolved by uniform distribution of AuNPs on the whole electrode surface.

Analytical signal also depends on the morphology of the electrode's surface. This is influenced by factors such as surface roughness. As a result, the amount of immobilized antibodies, their proper conjugation to the surface, and improved response signal depend on the amount and size of AuNPs deposited on the electrode surface. These are related to the deposition time and scan rate, and were optimized in Chap. 5.

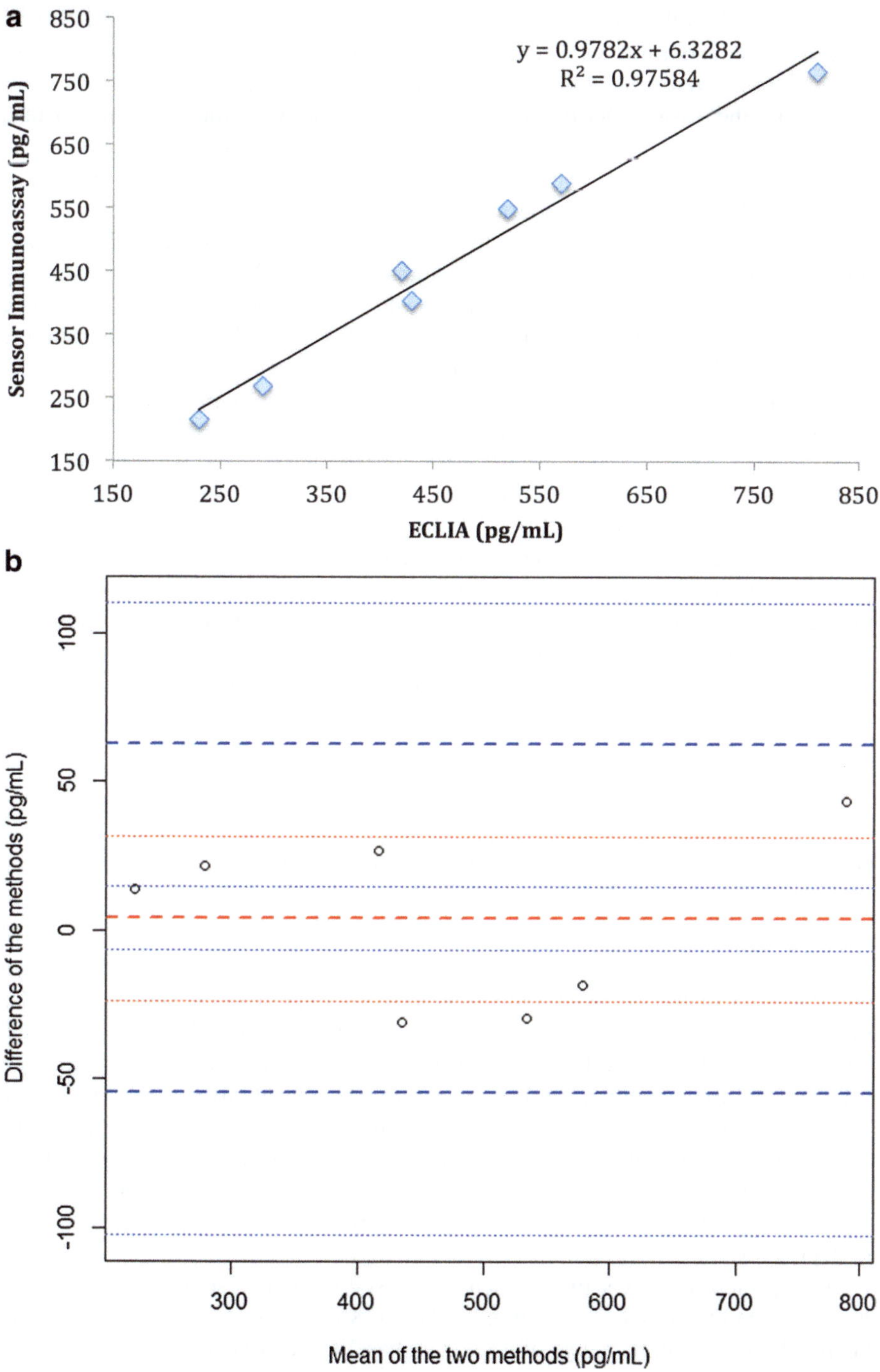

Fig. 6.15 Statistical analysis comparing ECLIA vs. CTX sensor immunoassay. (**a**) Correlation plot (**b**) Bland and Altman plot

Factors such as the concentration of GSH and Ab as well as the sulfo-NHS:EDC concentration and ratio were studied by a grid experiment where a series of concentrations were tested against each other to obtain the largest signal-to-noise ratio from the assay. Optimum GSH concentration was found to be 10 mM; 400 mM EDC: 100 mM sulfo-NHS: 10 μg/mL Ab ratio of 1:1:1 were the optimum concentration of the CTX antibody complex. As for Oc, similar concentration of GSH, sulfo-NHS, and Ab concentration was used.

The incubation time is another important parameter affecting the analytical performance of the immunoassay. At room temperature, the DPV response following the antibody complex immobilization improved with the increasing incubation time and then tended to constant values after 2.5 h. This showed the saturated binding both between the AuNPs and the antibody.

Similarly, in order to optimize the reaction time, the coupling procedure was repeated each time with a shorter reaction time until the signal change was too low to be detected. An identical procedure was repeated for a range of Ag concentrations in each case. The optimized reaction time, suggestive of saturated binding between the antibody and antigen, was 80 s.

As discussed in Chap. 4, temperature had no significant effect on the Ab immobilization and Ag-Ab reaction processes. Thus, to simplify the process, all the steps were performed at room temperature and the optimized incubation time of 2.5 h, as calculated in the same chapter, was used.

The highest activity of both Oc and CTX sensors was observed at pH 4.5, in which all the material (sulfo-NHS, EDC, and GSH as well as the antibody) are reported to have the best functionality [21, 22].

6.5 Specificity, Cross-Reactivity, and Cross-Talk

To investigate the specificity and cross-reactivity between analytes and Ab-functionalized surface, the interaction experiments were repeated with constant concentration of the opposite antigen along with PTH. The Oc sensor showed a minimum change from baseline value (<3%) with respect to CTX (100 ng/mL) and PTH (100 ng/mL) as interferents, indicating that the sensor did not interact with CTX or PTH to an appreciable degree over a wide concentration range, which can be attributed to efficient blocking of non-specific binding sites on gold (Fig. 6.16).

(**a**) Electrochemical response measured for Oc sensor incubated with CTX (100 ng/mL), PTH (100 ng/mL), and Oc (100 ng/mL), (**b**) Response measured for CTX sensor incubated with Oc (100 pg/mL), PTH (100 pg/mL), and CTX (100 pg/mL)

As expected, the CTX immunosensor showed obvious responses only towards its corresponding antigens. Apparently, the cross-reactivity at the array was again negligible (a minimum change of <3% from baseline value) (Fig. 6.16).

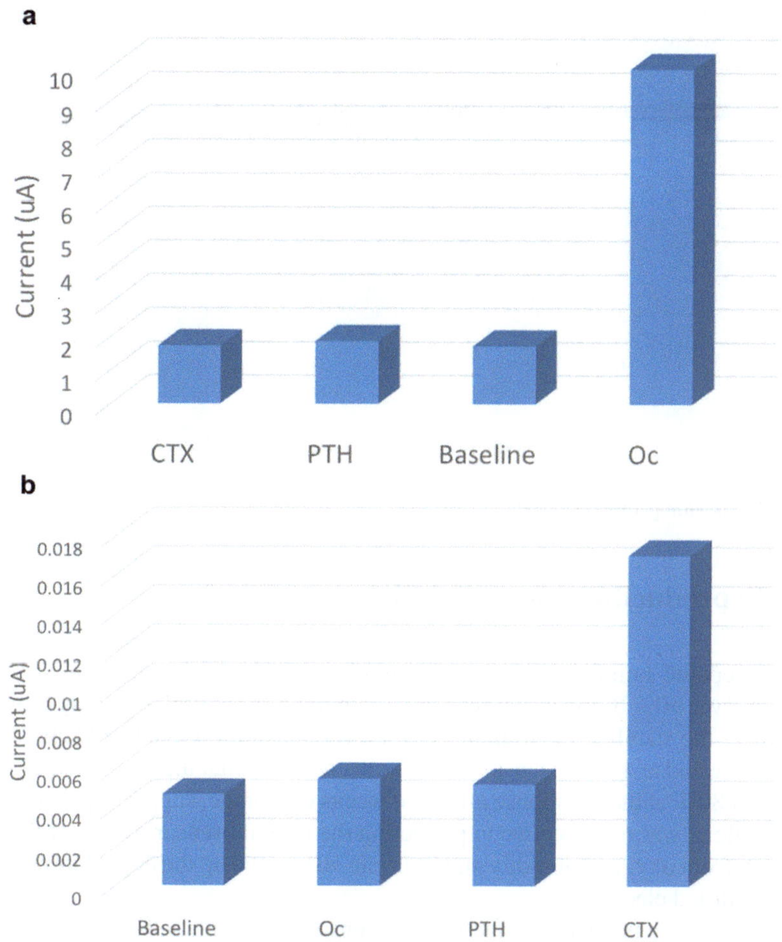

Fig. 6.16 (**a**) Electrochemical response measured for Oc sensor incubated with CTX (100 ng/mL), PTH (100 ng/mL), and Oc (100 ng/mL), (**b**) Response measured for CTX sensor incubated with Oc (100 pg/mL), PTH (100 pg/mL), and CTX (100 pg/mL)

An excellent immunosensor array must exclude cross-talk, which generally results from the diffusion of electroactive indicator produced on one electrode to the neighboring ones. In this work, as the Ab complex was immobilized on AuNPs, the Ab-antigen reaction would only accumulate on the AuNPs, resulting in a localized stripping reaction at the individual electrode. Hence, the possible cross-talk could be well avoided.

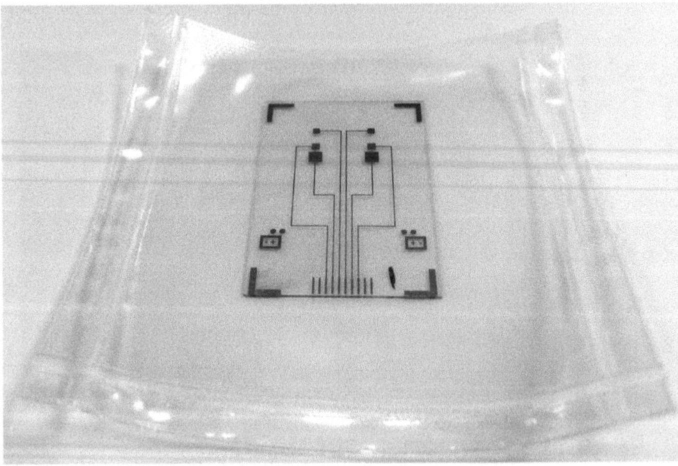

Fig. 6.17 Vacuum packed electrode

6.6 Reproducibility and Stability

The as-prepared immunosensor array could be stored in dry condition at 4 °C. In this way, over 90% of the initial responses remained for a week of storage. In order to improve the stability, the immunosensors were vacuum packed and kept at room temperature and at 4 °C (Fig. 6.17). The results revealed that the immunosensors lost more than 80% of their response after two weeks of storage, suggesting that vacuum packing alone was not effective in protecting the immunosensor over time. It seems that apart from oxygen, humidity and light can also speed up the degradation process of the modified electrodes.

The interassay precision of the immunosensor was examined based on repeated measurements of two different concentrations of CTX with five immunosensors. The coefficient of variation was 4.1%, indicating that the immunosensor had acceptable stability and reproducibility.

6.7 Conclusion

In this chapter, electrochemical detection method was applied in two different biosensor chips to detect serum levels of CTX and Oc. AuNPs were electrodeposited on gold electrodes to improve the surface area, provide better electron transfer kinetics, and higher background charging current. On the AuNP-modified gold electrode, CTX and Oc antibodies were covalently immobilized using cross-linkers, and different concentrations of the corresponding antigens were applied for direct determination.

Due to the excellent electrical properties of the as-prepared immunosensor, Oc sensor could detect serum levels of the marker within the range of 2.5–75 ng/mL. The reference value of Oc for adults of both genders is 9–42 ng/mL, suggesting that the sensor can acceptably detect Oc.

Similarly, CTX levels were successfully measured from 1 pg/mL to 1000 pg/mL. This is while the reference value of the marker is as follows:

- Males:
  - <18 years: not established
  - 18–30 years: 155–873 pg/mL
  - 31–50 years: 93–630 pg/mL
  - 51–70 years: 35–836 pg/mL
  - >70 years: not established.
- Females:
  - <18 years: not established
  - Premenopausal: 25–573 pg/mL
  - Postmenopausal: 104–1008 pg/mL.

This suggests that the sensor can acceptably detect CTX.

Compared to conventional methods, the proposed electrochemical sensors resulted in selective and sensitive measurement of each of the BTM markers with reduced interference reactions, reduced measurement time, need for less amount of reagent, and better stability in a simplified detection system.

The total assay time for these sensors is about 5 min (loading of antigen, incubation time, flushing with PBS, and testing). The good correlation between the results of ECLIA and that of these sensors showed the chips can be used in the clinically relevant range. Moreover, the fact that the initial calibration curve was plotted using pure antigen and the serum tested in the last step also contained other macromolecules suggests that the sensor is specific for the antigen and macromolecules present in serum do not affect the results. However, the fact that the Oc Ag used to plot our calibration curve was a full-length protein, while ECLIA measures N-mid Oc may have affected our results.

References

1. J. Brown, P.D. Delmas, L. Malaval, C. Edouard, M.C. Chapuy, P.J. Meunier, Serum bone gla-protein: a specific marker for bone formation in postmenopausal osteoporosis. Lancet **1**, 1091–1093 (1984)
2. K. Nimptsch, S. Hailer, S. Rohrmann, K. Gedrich, G. Wolfram, J. Linseisen, Determinants and correlates of serum undercarboxylated osteocalcin. Ann. Nutr. Metab. **51**(6), 563–570 (2007)
3. S. Atalay, A. Elci, H. Kayadibi, C. Onder, N. Aka, Diagnostic utility of osteocalcin, undercarboxylated osteocalcin, and alkaline phosphatase for osteoporosis in premenopausal and postmenopausal women. Ann. Lab. Med. **32**(1), 23–30 (2012)

4. P. Khashayar, H.A. Meybodi, G. Amoabediny, B. Larijani, Biochemical markers of bone turnover and their role in osteoporosis diagnosis: a narrative review. Recent Pat. Endocr. Metab. Immune Drug Discovery **9**(2), 79–89 (2015)
5. V. Jagtap, J. Ganu, N. Nagane, BMD and serum intact osteocalcin in postmenopausal osteoporosis women. Ind. J. Clin. Biochem. **26**(1), 70–73 (2011)
6. E. Chow, D. Hibberta, J. Gooding, Voltammetric detection of cadmium ions at glutathione-modified gold electrodes. Analyst **130**, 831–837 (2005)
7. M. Swartz, I. Krull, *Handbook of Analytical Validation* (CRC Press, Boca Raton, 2012)
8. G.L. Long, J.D. Winefordner, Linearization of electron capture detector response to strongly responding compounds. Anal. Chem. **55**(7), 713A–724A (1983)
9. J. Bartlett, C. Frost, Reliability, repeatability and reproducibility: analysis of measurement errors in continuous variables. Ultrasound Obstet. Gynecol. **31**, 466–475 (2008)
10. M. Mayer, J. yang, I. Gitlin, D.H. Gracias, G.M. Whitesides, Micropatterned agarose gels for stamping arrays of proteins and gradients of proteins. Proteomics **4**, 2366–2376 (2004)
11. A. Bernard, E. Delamarche, H. Schmid, B. Michel, H. Bosshard, H. Biebuyck, Printing patterns of proteins. Langmuir **14**, 2225 (1998)
12. P. Garnero, The contribution of collagen crosslinks to bone strength. BoneKEy Rep. **1**, 182 (2012)
13. S. Vasikaran, R. Eastell, O. Bruyere, A. Foldes, P. Garnero, A. Griesmacher, M. McClung, H.A. Morris, S. Silverman, T. Trenti, D. Wahl, C. Cooper, J. Kanis, for the IOF-IFCC Bone Marker Standards Working Group, Markers of bone turnover for the prediction of fracture risk and monitoring of osteoporosis treatment: a need for international reference standards. Osteoporos. Int. **22**, 391–420 (2011)
14. S. Chubb, Measurement of c-terminal telopeptide of type I collagen (CTX) in serum. Clin. Biochem. **45**(12), 928–935 (2012)
15. S. Baim, P.D. Miller, Perspective: assessing the clinical utility of serum ctx in postmenopausal osteoporosis and its use in predicting risk of osteonecrosis of the jaw. J. Bone Miner. Res. **24**, 561–574 (2009)
16. M. Bonde, P. Garnero, C. Fledelius, P. Qvist, P. Delmas, C. Christiansen, Measurement of bone degradation products in serum using antibodies reactive with an isomerized form of an 8 amino acid sequence of the c-telopeptide of type I collagen. J. Bone Miner. Res. **12**(7), 1028–1034 (1997)
17. C. Rosenquist, C. Fledelius, S. Christgau, B. Pedersen, M. Bonde, P. Qvist, C. Christiansen, Serum crosslaps one step ELISA. First application of monoclonal antibodies for measurement in serum of bone-related degradation products from c-terminal telopeptides of type I collagen. Clin. Chem. **44**(11), 2281–2289 (1998)
18. Y.-H. Yun, A. Bhattacharya, N.B. Watts, M.J. Schulz, A label-free electronic biosensor for detection of bone turnover markers. Sensors **9**(10), 7957–7969 (2009)
19. M. Ramanathan, M. Patil, R. Epur, Y. Yun, V. Shanov, M. Schulz, W.R. Heineman, M. Datta, P. Kumta, Gold-coated carbon nanotube electrode arrays: immunosensors for impedimetric detection of bone biomarkers. Biosens. Bioelectron. **77**, 580–588 (2016)
20. A. Killard, B. Deasy, R. O'Kennedy, M. Smyth, Antibodies: production, functions and applications in biosensors. Trends Anal. Chem. **14**(6), 257–266 (1995)
21. M. Fischer, Amine coupling through EDC/NHS: a practical approach. Methods Mol. Biol. **627**, 55–73 (2010)
22. M. Hormozi-Nezhada, E. Seyedhosseinia, E. Robatjazia, Spectrophotometric determination of glutathione and cysteine based on aggregation of colloidal gold nanoparticles. Sci. Iran. **19**(3), 958–963 (2012)
23. T.R. Holford, F. Davis, S.P. Higson, Recent trends in anti- body based sensors. Biosens. Bioelectron. **34**(1), 12–24 (2012)
24. K. Omidfar, F. Khorsand, M. Darziani Azizi, New analytical applications of gold nanoparticles as label in ab based sensors. Biosens. Bioelectron. **43**, 336–347 (2013)
25. E. Katchalski-Katzir, D. Kraemer, C. Eupergit, A carrier for immobilization of enzymes of industrial potential. J. Mol. Catalysis B Enzymatic **10**, 157–176 (2000)

26. F. Rusmini, Z. Zhong, J. Feijen, Protein immobilization strategies for protein biochips. Biomacromolecules **8**, 1775 (2007)
27. J. La Belle, A. Fairchild, U. Demirok, A. Verma, Method for fabrication and verification of conjugated nanoparticle-ab tuning elements for multiplexed electrochemical biosensors. Methods **61**(1), 39–51 (2013)
28. W. Vastarella, L. Della Seta, A. Masci, J. Maly, M. De Leo, L. Maria Moretto, R. PIlloton, Biosensors based on gold nanoelectrode ensembles and screenprinted electrodes. Int. J. Environ. Anal. Chem. **87**(10–11), 701–714 (2007)
29. P. Khashayar, G. Amoabediny, M. Hosseini, R. Verplancke, F. Razi, J. Vanfleteren, B. Larijani, An electrochemical biosensor based on AuNP-modified gold electrodes for selective determination of serum levels of osteocalcin. Sensors **17**(11), 3367–3374 (2017)

Chapter 7
Osteokit Fabrication

Abstract As nowadays, microfluidic technologies are widely employed in developing point-of-care diagnostics to address global health issues because of their potential advantages of low sample and reagent consumption, high throughput and sensitivity, large surface-to-volume ratio, and other benefits related to miniaturization, we decided to integrate our electrochemical chip into a microfluidic system. The fabrication of microfluidic channels is commonly costly and requires laboratory-intensive cleaning, photolithography, and etching or baking steps in cleanroom environments, making it difficult to modify. Besides, proper channel enclosure without deforming small features or without clogging of the channel during the bonding process is challenging. We therefore developed a cheap, reliable, and rapid method for the fabrication of microfluidic channels using double-sided tapes, enabling not only highly uniform cross-sectional dimensions along the channels but also proper adhesion in hybrid systems, composed of different layers.

7.1 Introduction

Nowadays, considering the limited resources, the use of rapid, cheap, simple-to-operate, disposable, and accurate PoC diagnostics is of great importance and can have a critical impact on clinical decision-making. Although advanced diagnostic technologies such as polymerase chain reaction (PCR) and ELISA are widely used, their application in developing countries is limited because of the limited availability of laboratory infrastructure, skilled personnel, and financial supports [1].

This chapter contains information published in the article from the author and her team in "Micromachines" [28].

As a result, microfluidic technologies have been widely employed in developing PoC diagnostics and to miniaturize immunoassays to address global health issues because of their potential advantages of low sample and reagent consumption, high throughput and sensitivity, large surface-to-volume ratio, and other benefits related to miniaturization [2]. However, as mentioned in Chap 3, most microfluidic devices require both complex fabrication processes and external instruments [3]. Several aspects need to be considered while designing a microfluidic device, such as selection of materials, dimensions of the microfluidic devices, and fluidic control devices (such as pumps, valves, and mixers).

P. Khashayar, *Point-of-Care Solution for Osteoporosis Management*,
DOI 10.1007/978-3-319-55053-4_7

Some of the materials used for fabrication of microfluidic devices for PoC diagnostics include paper (nitrocellulose or cellulose), glass, silicon, plastics (cyclic olefincopolymers or polystyrene), and polycarbonate. Currently, polymers such as PDMS and thermoplastics are most frequently applied to facilitate the translation of microscale systems from simple laboratory tools to commercially marketable products [4]. The latter are more favorable as they allow for the production of dimensionally stable and mechanically robust components while the feature fidelity and integrity are easier to maintain during replication processes [5]. Among thermoplastics, cyclic olefin copolymer (COC), poly (methyl methacrylate) (PMMA), polystyrene, and polycarbonate are frequently used for fabricating microfluidic systems.

Despite the above-mentioned advantages, the fabrication of microfluidic channels using these materials requires costly and laboratory-intensive cleaning, photolithography, and etching or baking steps in cleanroom environments. Thus, design iterations to modify the device are labor and resource-intensive [6]. Besides, proper channel enclosure without deforming small features or without clogging of the channel during the bonding process is challenging [7, 8]. Nowadays, the key step to the final integration includes plasma or UV/ozone treatment, thermal fusion, microwave welding, adhesives, surface modification or solvent bonding, which apart from being costly and time-consuming, does not always provide acceptable results [9–11].

Hot embossing and thermal bonding are currently the methods of choice for fabricating and sealing microfluidic channels in thermoplastic materials [12]. These methods allow a relatively fast and easy fabrication of microfluidic devices but they generally need expensive equipment and a cleanroom environment [13]. The temperature and pressure applied during these steps also limit the variability of the materials used.

Hot embossing is a popular method for fabricating microfluidic channels in thermoplastic materials. The fabrication process typically starts with a PDMS or epoxy-based master mold, which is made using well-established photolithographic processes [14]. Shrinkage of this master mold and subsequent misalignment are quite common with large, centimeter-order devices [15]. The quality of microfluidic channels during the replication process is also of great importance. During thermal fusion, the polymer is heated to or slightly above its glass transition temperature (Tg), and thus becomes soft and sensitive to heterogeneities in load distribution during the bonding process.

In this chapter we report the design and fabrication of a sensitive multiplexed microfluidic proteomic platform integrated with electrochemical sensing that can easily be translated into a protein biomarker diagnostic. This chapter contains the information published by the author and her team in Journal of the "Brazilian Society of Mechanical Sciences and Engineering" [16]. To our knowledge this is the first such device designed as a step towards developing a PoC system for the measurement of several BTMs.

7.2 System Design and Fabrication

7.2.1 Design and Fabrication of Sensor Chip

The electrochemical chip was the glass-based gold electrode fabricated in Chap. 5. In brief, it consisted of six gold electrodes arranged in a 2 by 3 electrode arrays on a glass. The glass was 25.5×85.5 mm^2, leaving a 25.5×10 mm^2 uncovered area, after assembling with the microfluidic platform, for the contact pads to be connected to the read-out circuit using a custom-made PCB. Each gold working electrode (1.5×1.5 mm^2) was placed between one gold pseudo-reference (1.4×1.4 mm^2) and one gold counter electrode (2.5×2.5 mm^2), as depicted in Fig. 7.1.

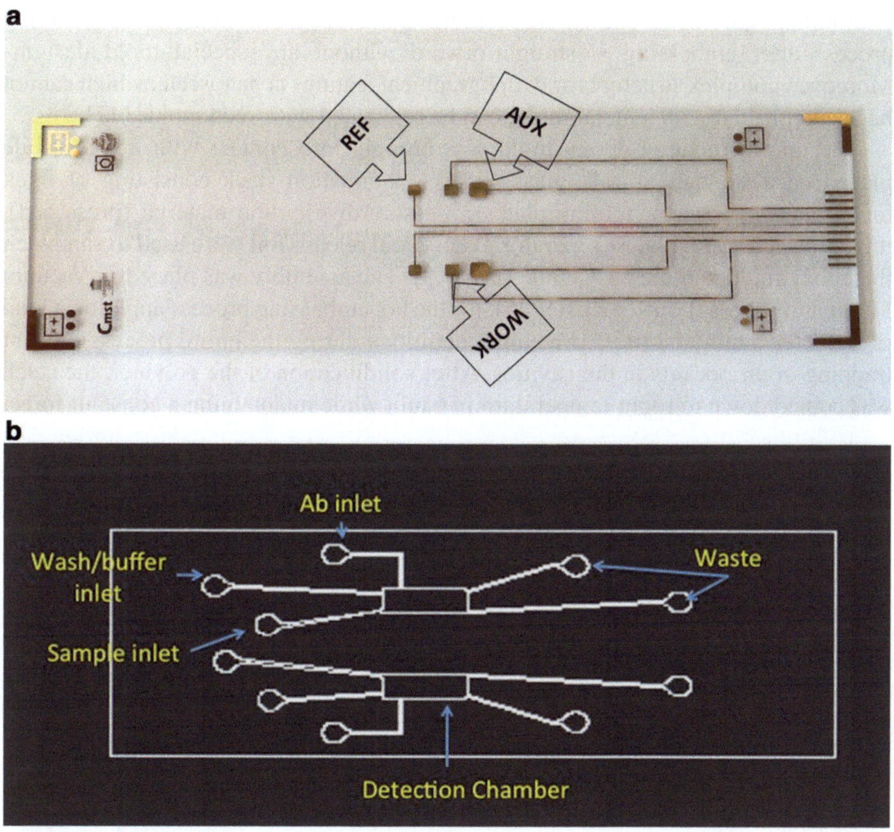

Fig. 7.1 (**a**) Schematics of the electrochemical sensor chip. (**b**) Structure of the microfluidic manifold

7.2.2 *Design and Fabrication of Microfluidic Manifold*

In order to develop a fully operational electrochemical diagnostic prototype, the microfluidic platform consisted of the microchannels and a chamber necessary to expose electrodes in the electrochemical sensor chip to the sample, surface modification solution, and detection probe. All delivered samples and reagents were routed to the waste outlets.

For the fabrication of the microfluidic channel layer, the design of the microfluidic channels was first drawn in a computer aided design and drafting software (Autodesk, USA). The design was then used to fabricate epoxy molds. Figure 7.2 illustrates the process used to fabricate epoxy (FR-1080)[1] mold from PDMS (Sylgard 184) stamps. The PDMS stamps were themselves fabricated using soft lithography using SU-8 on a silicon wafer, which was itself structured using photolithography. The use of these molds has several advantages. De-molding process after embossing is straightforward without any special mold designs. Moreover, complex structures and topographical features at nanoscale, which cannot be accomplished with a metal mold, can be replicated with such molds [17].

The microstructured epoxy mold was brought into contact with a COC slide placed on a standard 4″×4″ glass slide. A lamination stack consisting of thick metal plates (carrier or tooling plates), pressure distribution material (press pad), thin, polished metal plates (separator plates), and release foil were used to sandwich the epoxy mold and the COC slide (Fig. 7.3). The assembly was placed in Vacuum Laminating Press (Lauffer RLKV 25), for the hot embossing process applying a load of 3 kN for 5 min at 110 °C. Vacuum conditions during the entire process prevent trapping of air pockets in the cavities. After solidification of the polymer, the stack was cooled down to room temperature in 6 min while maintaining a constant force.

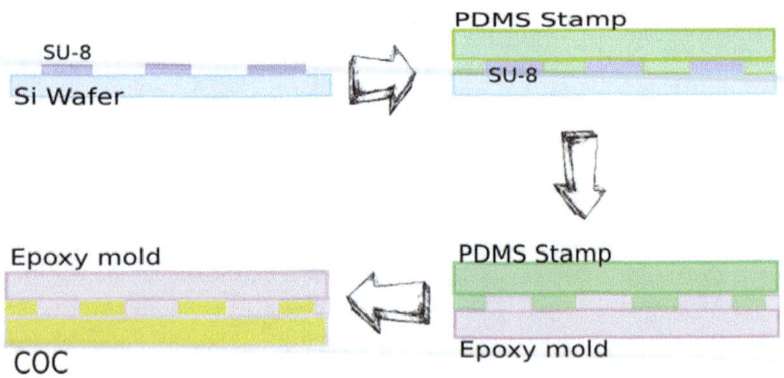

Fig. 7.2 Fabrication process of epoxy molds

[1]This is the main type of epoxy that is generally used for hot embossing applications (up to 135–140 °C).

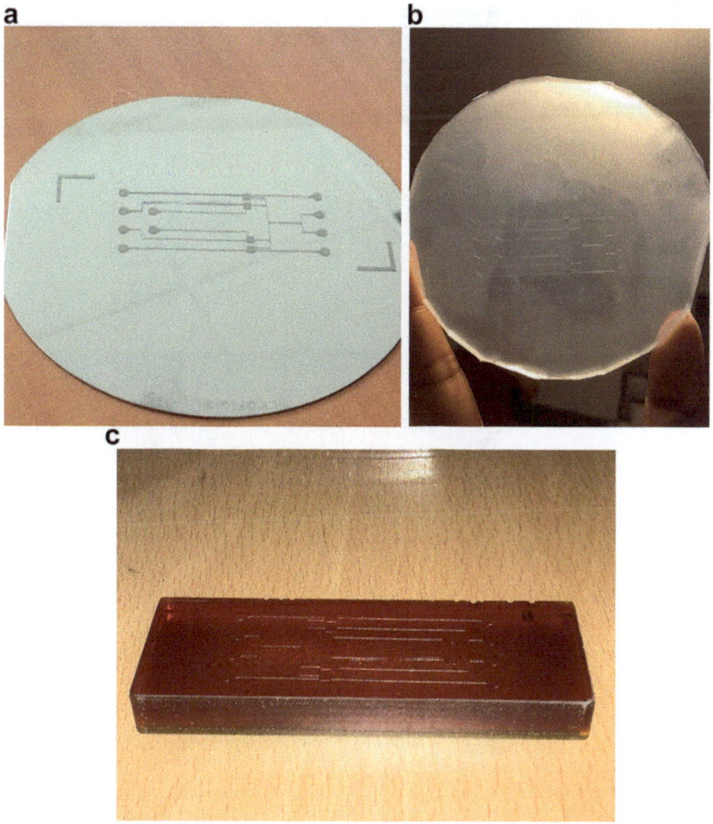

Fig. 7.3 (**a**) Si wafer used to fabricate the PDMS stamp. (**b**) PDMS with negative copy of SU-8 master. (**c**) Epoxy mold

Then the assembly was unloaded and the epoxy mold was carefully peeled off from the embossed COC slide to avoid tearing or ripping of the epoxy mold.

As the next step, thermal bonding with Vacuum Laminating Press was used to enclose the open structure of the COC slide replica with a glass slide which also acted as the electrochemical chip (as mentioned in Chap. 6). In this regard, the stack was loaded into the press, and bonded at 3 kN and 110 °C for 10 min and then were cooled to room temperature under the same pressure; it was, thereafter, unloaded. The process took 20 min. It was associated with several limitations (Fig. 7.4).

- Some deformation was noted in the fabricated microstructures in the COC.
- The process was time-consuming and the stack assembly was heavy.

Therefore, the microfluidic channels were realized using a double-sided medical grade tape (3M) of 86 μm thickness, which was previously laser machined to generate the microchannel structures. The CAD design was used to define the cutting path through the double-sided tape, using a laser beam. Therefore, any a

b

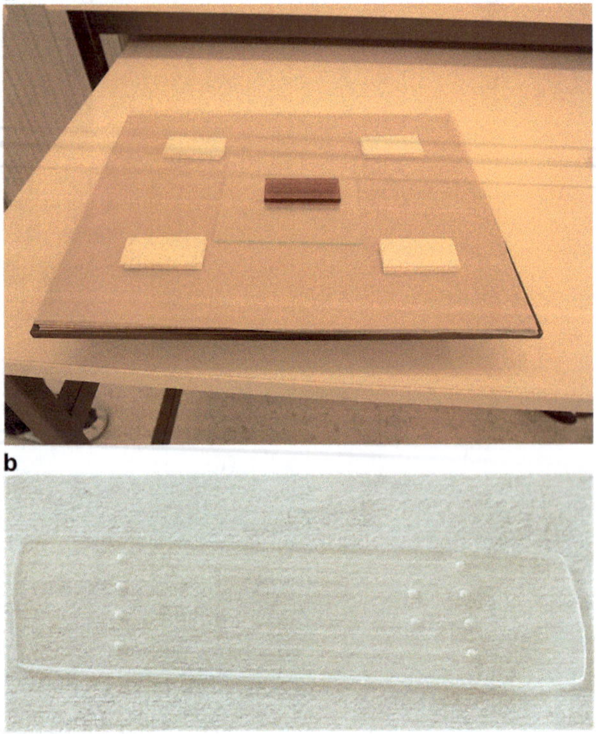

Fig. 7.4 (**a**) Lamination stack (partially shown). (**b**) Deformed COC after the embossing process

modification of the design is straightforward and easy to implement. Here, a focused 9.6-μm wavelength CO_2 laser beam was used to partially cut through the tape at predefined locations to ease handling, aligning, and assembling in general. Figure 7.5 shows the cross-section of a partially cut tape. It should be noted that the laser beam spot has a certain diameter (approximately 100 μm), which we compensated for in the CAD design. The patterned tape was then used to integrate the microfluidic manifold, i.e., the COC slide, with the glass sensor chip and aligned using a manual mask aligner (SET MG1410). Using this technique, patterning of COC was not required anymore.

An exploded-view drawing of the three-layer microfluidic manifold with the sensor chip is presented in Fig. 7.6. In this regard, the sensor chip (gold electrodes on the glass) acted as the bottom layer. The middle layer (tape) defined the capture/reaction chamber and channels. The topmost layer in the chip serves as the covering for the channels. COC was used for this layer as drilling the I/O holes was much easier in it compared with glass. This layer is a rectangular piece of COC (COC, Topas 8007) (25.5 mm by 75.5 mm, 1 mm thickness) drilled with 1 mm diameter holes that serve as the reservoirs to introduce the sample and buffer solutions as well as to collect the waste. The fluid flows in the microfluidic channels

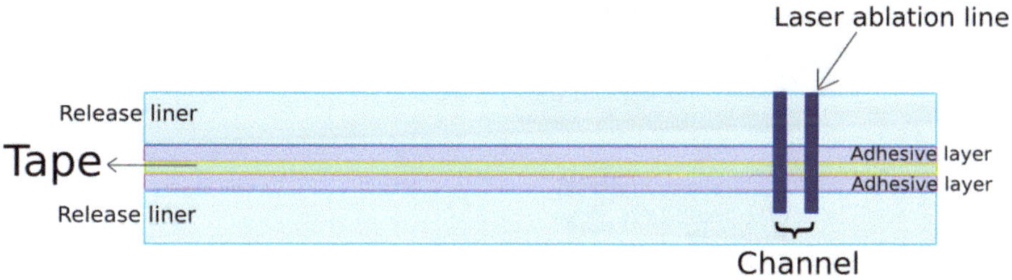

Fig. 7.5 Tape was partially cut using laser ablation to facilitate the handling, aligning and assembling process

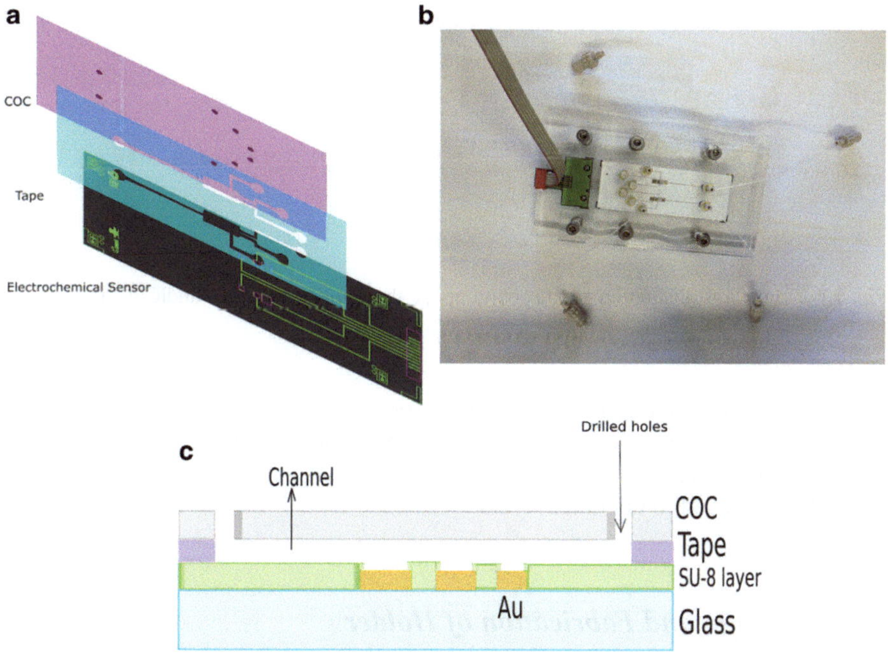

Fig. 7.6 Diagram of the microsystem device (**a**) CAD design of each layer, (**b**) Microfluidic device positioned in between 2 PMMA plates, acting as microfluidic chip holder, (**c**) Cross-section of the microfluidic device

and reaction occurs in the reaction cells where the fluid is exposed to the electrodes on the bottom layer. Figure 7.7 explicitly shows the cross-sectional view of each layer in the chip.

The channels had a width of 200 μm; this is a safe value considering that our studies have shown that 100-μm-channels are wide enough for serum/blood samples to pass through without clogging or sedimentation. The capture/reaction chamber had a width and length of 2.9 and 8.9 mm. The height of the channels and chambers were equal to the thickness of the tape (86 μm). Transparency of the injected fluids

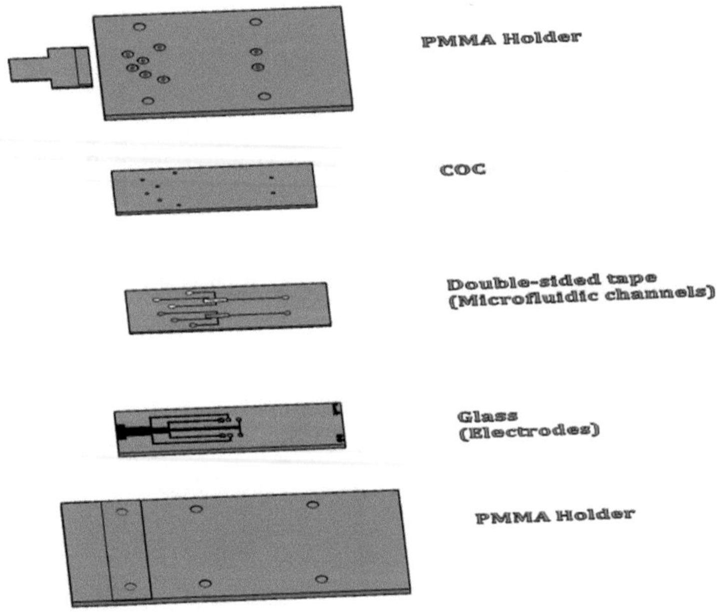

Fig. 7.7 Schematic illustration of the build-up of the multilayer microfluidic device and the microfluidic chip holder

hindered the observation of fluid flow; therefore a blue dye was used to characterize fluid flow profiles in the capture/reaction chamber and channels. In this approach, each layer is processed as if they were independent substrates. They were then aligned and bonded to the other layers.

7.2.3 Design and Fabrication of Holder

In order to interface the microfluidic chip with external fluidic and electrical connections, a custom-made holder was designed and fabricated in-house by micro-milling in a 10 mm PMMA sheet and cut out in the dimensions of 65×126 mm^2. The docking station, used to house the chip, functioned as a base with female ports fully integrated into the device to reduce cost and improve reproducibility and reliability of the results [18].

The top PMMA plate includes circular, drilled holes of 1 mm diameter (centered relative to the I/Os) that fit with the corresponding connection ports in the COC. The ports for incorporating fluidic connectors, purchased from LabSmith, were glued to these PMMA holes. These interconnects were chosen because of their small footprint that allowed a high density of fluidic I/O ports and minimal dead volume.

A circular notch of approximately 0.3 mm depth was designed on the inner surface of this plate to place the Viton O-rings (2 mm inner diameter, 4 mm outer diameter, 1 mm thickness), again centered relative to the inlets/outlets. The O-rings were used to provide a leak-free interface between the holder and the inlets and outlets on the COC. Tygon tubing was used to introduce the fluids from a 1 mL Luer-Lok tip, insulin syringe (BD, Franklin Lakes, US). The lower half of the holder was completely featureless. The two sides were tightened together with screws after the assembled device was put in between and aligned by eye.

7.2.4 Assembling and Sealing

In order to minimize surface contamination, parts were blown off with high-pressure nitrogen before the assembling step. The assembly of the layers started with removing one of the release liners and bonding the micro-patterned adhesive tape to the COC substrate. A roller was used to avoid the entrapment of air at the interface. Thereafter, the other release liner was removed, and the micro-positioning table and the microscope of a mask aligner system were used to align the COC/tape stack with the glass substrate before bringing them into contact and creating a tight bond. It should be noted that considering the design, the second part can also be performed manually. This process of stacking—layer, tape, layer, tape—can be repeated as needed to complete the microfluidic device. For the final bonding, the entire assembly was pressed together manually. The microfluidic device was then placed in the custom-made microfluidic chip holder (Fig. 7.6).

7.3 Microfluidic Characteristics

The improvement of detection sensitivity and analysis time has been a key motivation for the development of analytical microsystems [19, 20]. Size reduction lowers the consumed reagent and sample volume and also results in portability. It, however, may compromise the detection limit of the device, especially for low flow-rates [21]. This is while the maximum allowed pressure for reliable operation and integrity of the microfluidic device limits the flow-rate itself [22]. Therefore, the parameters that contribute to the sensitivity and specificity of the measurements in such devices are as follows:

- Successful and functional immobilization of the antibodies on sensor surfaces to perform reagentless detection
- Amplifying reaction signals
- Flawless flow of the solutions in the microchannels during preparation steps and electrochemical process
- Absence of cross-talk due to the diffusion of electroactive products from one electrode to a neighboring one causing interference.

7.3.1 *Channel Characterization*

The dimensions of the resulting microfluidic channels along with the characteristics of the material used to make the microfluidic channels can affect the fluid flow.

7.3.1.1 Channel Dimension

The narrowest channel, which could be fabricated using the laser-cutting technique, has a width of around 100 µm, because of the laser's beam spot. Therefore the technology is not applicable everywhere, but offers a cost-effective, simple, and accurate solution for several applications.

Figure 7.8 presents the difference in the width of the microfluidic channels seen from the front- and back-side of the tape before bonding. The trapezoidal shape of the cross-section is typical for microfluidic channels fabricated using laser-cutting and is due to the laser-cutting and can be contributed to the focal plane of the laser beam (Fig. 7.9). Due to the trapezoidal cross-section, the width of the microfluidic channels varied from to $110 \pm 5\,\mu m$ at the top to $65 \pm 5\,\mu m$ at the base of the microfluidic channels.

In order to quantify the uniformity of the cross-sectional geometry of the microfluidic channels along their length, a microfluidic channel was diced along its length at 5 mm intervals and assessed with an optical microscope (Nikon Eclipse LV100) and non-contact optical profiler (WYKO NT3300) to assess its width and height, respectively (Fig. 7.10).[2] The low relative standard deviations confirm the uniformity of the channel dimensions.

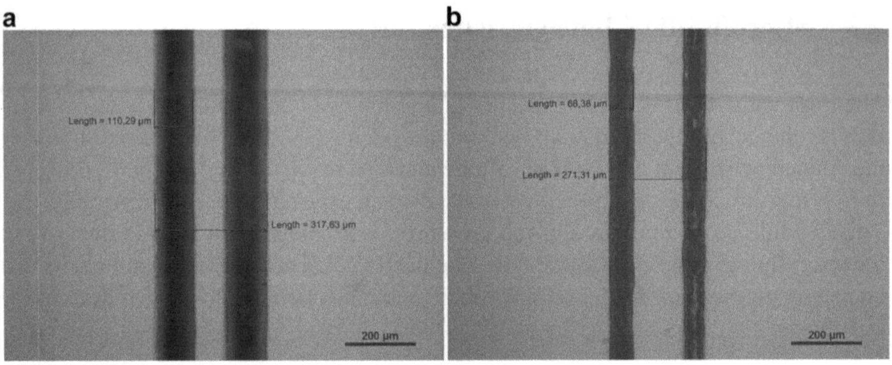

Fig. 7.8 Optical microscope images of the width of a 100 µm wide microchannel before bonding seen from the (**a**) back and (**b**) front side of the tape, showing the trapezoidal shape of the microfluidic channels. The *dark lines* represent the ablated tape

[2]As dicing the Glass–Tape–COC at 5 mm intervals was troublesome, COC–Tape–COC was used instead.

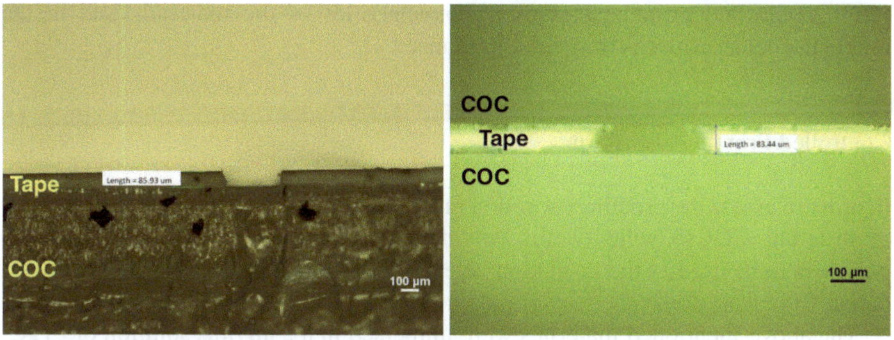

Fig. 7.9 Optical microscope images of the cross-section of a microfluidic channel, showing its height (**a**) before (COC–Tape) and (**b**) after bonding (COC–Tape–COC)

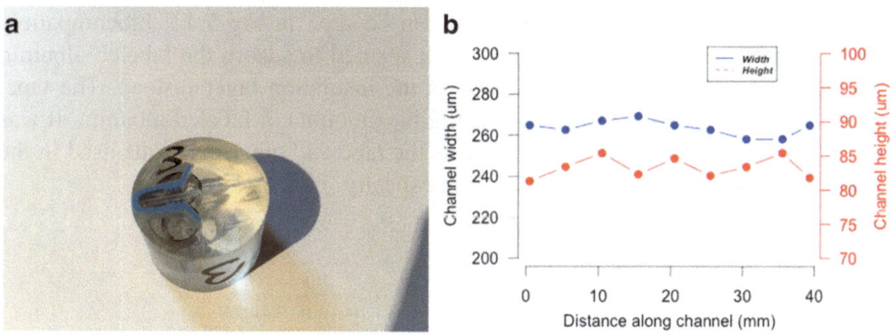

Fig. 7.10 (**a**) Cross-section of the chip, prepared by dicing along the length of the microfluidic channel at 5 mm intervals, (**b**) Channel height and width at 5 mm intervals along the length of a 250 μm wide microfluidic channel, measured using an optical microscope and non-contact optical profiler

7.3.1.2 Channel Structure

The typical permeability of COC for oxygen is much less than the value observed for other materials popular for the preparation of microfluidic chips [23]. This material property is of great importance as the presence of oxygen in the channel would disturb the reactions and measurements in different ways.

As mentioned earlier, the material used to fabricate microchannels should provide proper channel enclosure without deforming small features or clogging the channel during the bonding process [7, 8]. To overcome these concerns, we used double-sided tape both as glue to bond layers and as channel structures.

One of the main purposes for the fabrication of microfluidic devices with feature size in the micrometer range is to reduce the quantity of the needed sample. In this regard, non-specific adsorption of proteins to channel walls and cross-talk to other channels should be prohibited as it may cause unpredictable reduction in the concentration of the assay component and slow its transport rate. As a result, controlled protein adsorption characteristics of the material used in the microchannel structure is of great importance.

Controlled Protein Adsorption

In order to assess the protein adsorption rate in microchannels, the auto fluorescence of different COCs/Cyclic Olefin Polymers (COP), the double-sided tape, and polymers to be used as the insulation layer on the immunoelectrode were measured (Chap. 5).

The above-mentioned materials were immersed in a 2 mg/mL solution of FITC-Albumin in 0.05 M PBS (pH 7.0) for 2 h and subsequently rinsed with DI water [24]. Thereafter the amount of their protein adsorption capacity was determined by measuring fluorescence ($\lambda_{excit} = 495$ nm, $\lambda_{emis} = 520$ nm) in the labeled materials using a fluorescence spectrometer. As it can be seen in Fig. 7.11, biocompatible polyimide (PI) was the only material which seemed to adsorb the labeled albumin and thus biocompatible SU-8 was used as the insulation layer instead. The tape, on the other hand, seemed to adsorb negligible amounts of labeled albumin. It was concluded that cross-talk is not an issue as the tape and other materials used in the Osteokit had negligible protein adsorption capacity.

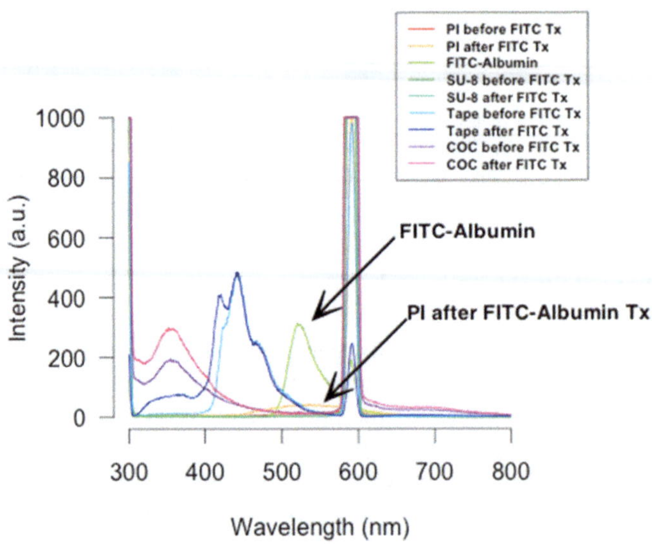

Fig. 7.11 Fluorescence emission ratio of different materials to be used in channel fabrication before and after treatment with FITC

7.3.1.3 Capture/Reaction Chamber characterization

In general, the detection system in the PoC microfluidic immunoassay devices must have high sensitivity and short response time [25]. There are several parameters that contribute to the sensitivity of the capture/reaction step. These parameters include the area of the electrode surface, flow-rate of sample and wash into the capture/reaction chamber, and surface modification step.

Non-optical detection, especially the electrochemical techniques are important for the microfluidic immunoassay as they are highly sensitive, rapid, and can be easily integrated by micromachining technology [26]. However, the major disadvantage of this method is the background noise. As mentioned earlier, AuNPs were used for electrode modification because of their excellent acceleration of electron transfer and thus signal amplification and low background noise. It was therefore concluded that AuNPs on the electrodes acted as a powerful multifunctional layer that improved in situ immobilization of antibodies on the electrodes, amplified the measured signals, and improved the sensitivity of this electrochemical sensor.

With these parameters in mind, we fabricated a $2.9 \times 8.9\,\mathrm{mm}^2$ capture/reaction chamber that was $86\,\mu\mathrm{m}$ deep (volume $= 2.2\,\mathrm{mm}^3$) to maximize the number of interactions between the agents needed to functionalize the surface as well as that between target analytes and the functionalized surface.

The special design of the channels along with the application of the chronoamperometry technique to reduce the measurement time and the electrolysis risk also helped overcome the bubble formation concern. It should be noted that in the liquid environment, bubbles are easily generated by electrolysis, and influence the result.

7.3.2 Flow Characterization

For characterizing the fluid flow in the capture/reaction chamber, a colored dye was utilized to visualize the flow of the fluids entering and exiting the capture/reaction chamber. Injecting a colored dye demonstrated the capability of the device to move fluids both through the channels and across different layers over the detection zones. According to our design, the fluid is distributed from an open inlet on the top COC substrate of the device into a single channel, which is connected to an open outlet on the same layer. During this path, the fluid passes over the detection zone, which is located on the bottom glass substrate. If various ports are open, the injected sample can be distributed through different channels to the outlets. This ability to redirect the sample into multiple detection zones makes it possible for the microfluidic device to be used for simultaneous measurement of multiple markers in a sample. No leakage was noticed in between adjacent channels.

Transporting liquids from the inlet to the reaction chamber and to the outlet is a key step in the microfluidic immunoassay device. In this project, the fluidic control of the analyte in and out of this chamber was achieved through capillary motion and pressure driven flow, generated by a syringe pump. The flow behavior mainly depends on the inlet flow-rate and thus we tried to introduce liquid into the system with a constant volumetric flow-rate. Flow-rates higher than 3500 μL/min were observed to be associated with gas bubble introduction into the channels and so were avoided.

Moreover, a wash step was performed at the beginning to ensure the transfer of remaining air in the interconnects, channels, and chambers to the outlet. This is because when hydraulic pressure is applied, gas bubbles remaining in the interconnection holes can be trapped in the reaction region during the detection process and cause significant problems by disturbing fluid flow or drying electrodes-on-chip [27].

Liquid movement is also controlled by different fluidic resistance within the parallel channel network. During the experiment, most of the pressurized solution flowed into the reaction chamber because the fluidic resistance of other channels, which were filled with air at this point, was higher than that of the reaction chamber. The adopted design was also accountable for reduced back-flow and improved flow of the solutions without dead volume formation. This was achieved by computational fluid dynamic (CFD) simulation and testing different designs in the lab.

7.3.3 CFD

All numerical computation presented in this work were performed using COMSOL Multiphysics (Finite Element Method) and Fluent (Finite Difference Method). Simulation was performed for a two-dimensional, incompressible Navier–Stokes flow. A no-slip condition was assumed along the walls of the microchannel and physical values of water (dynamic viscosity: 1.002e−3 cP; density: 1000 kg/m^3) were used in the numerical computation. The effect of the position of the inlet and outlet microchannels and the geometry of the recognition chamber on the flow profile were investigated.

The tests were performed in two states:

- Steady state, where the injection time is estimated to be infinite. This profile studies the velocity and pressure at different points of the microchannel. Both outlets are considered open all through the experiments, whereas only one of the inlets is open in each set of tests. As a result, the pressure at the outlet is equal to one atmosphere.
- Transient state, which aims to study the flow profile in the system. In this simulation, the channels are considered to be filled with air before the first injection. A constant flow-rate (400 μL/min) is applied to the inlet, pushing the air out of the outlet.

In order to validate the software setup, a single-channel system was used. The pressure drop in the channel is shown in Fig. 7.12. As it could be seen, the pressure

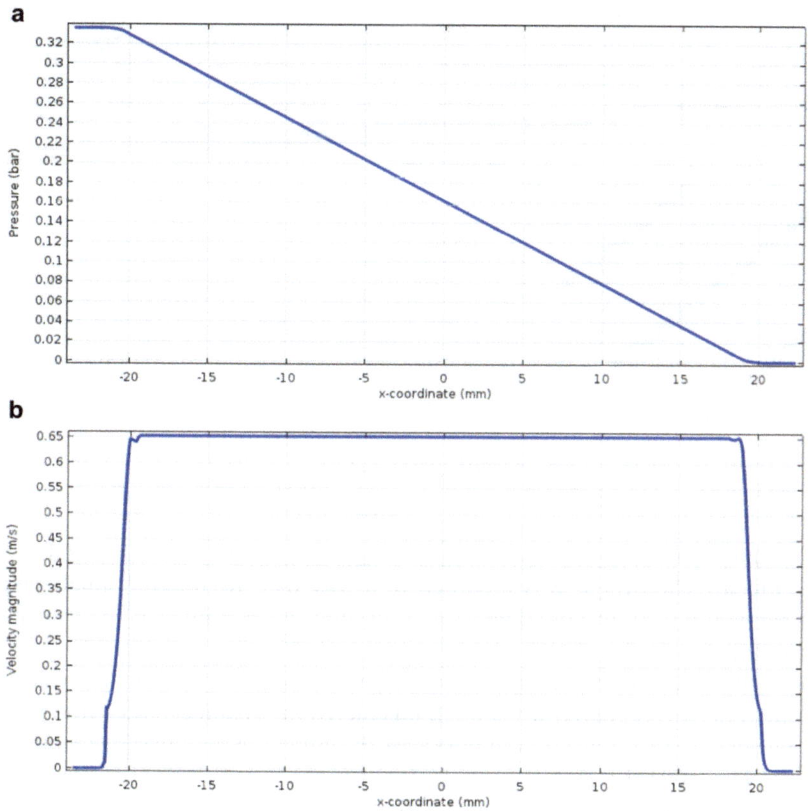

Fig. 7.12 (**a**) Pressure and (**b**) Velocity change through single microchannel

at the beginning and end of the microchannel is constant and the pressure drop through the channel follows a linear drop. This could be explained by the small size of the fully developed length.

The velocity change through the channel is shown in Fig. 7.12. In the inlet, the velocity was minimum and the pressure was maximum. The considerable velocity increase at the beginning of the channel is due to the mass conservation law and the fact that the fluid enters a small channel with a smaller surface area compared to the inlet. Thereafter the velocity remains constant until it reaches the outlet, where a considerable decrease is noted.

When studying the velocity in the microchannel's cross-section, in view of the no-slip condition, the velocity along the walls was observed to be zero and the maximum velocity was reported in the center of the microchannel (Fig. 7.13). The maximum shear stress, on the other hand, was noted on the walls and the minimum in the center of the channel.

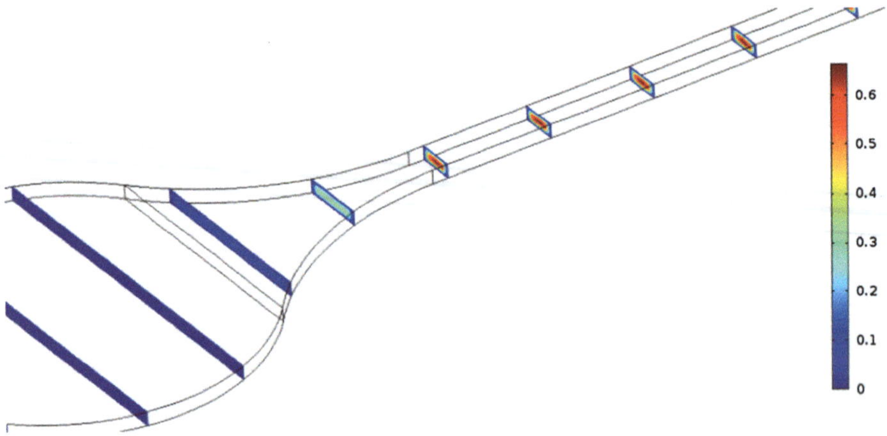

Fig. 7.13 Velocity contours in the channel's cross-section

The multi-channel network setup was drawn in Gambit software and studied using Fluent. The design was modified in a way that the microchannels and the chamber would be filled completely, the back-flow to other inlets is reduced and no dead volume is noted.

As mentioned earlier, maximum pressure was noted in the inlet. A considerable pressure drop was observed when the fluid entered the chamber, where the pressure became equal to that of the closed inlets. As the outlet pressure was one atmosphere and the pressure in the chamber was higher than the atmosphere pressure, the fluid was guided through the outlet with a negligible back-flow.

It should be noted that in the course of the simulation, dead volume was identified in the square chamber along with the inlets and outlets, suggesting that the immobilization and washing of the electrodes would not be effective. Based on these observations, a design modification was incorporated, where an additional outlet was added and the angels of the inlets and outlets were changed so that the dead volume was eliminated, the back-flow in other inlets was negligible and the fluid flow trajectory, the velocity field, and pressure were evenly distributed across the chamber (Fig. 7.14).

Figure 7.15 represents the fluid flow trajectories in the final optimized design. The results showed that the chamber was completely filled after the first second, and thereafter the volume fraction of the fluid remained constant. The system was completely filled after a longer injection time.

The maximum Reynolds number was computed to be less than 0.01, confirming laminar flow in the microchannels except for the corners (Fig. 7.16). The reduced shear stress and the turbulent flow observed at the corners of the chamber increased the risk of possible sedimentation, pointing out the need for gentle washing after each step.

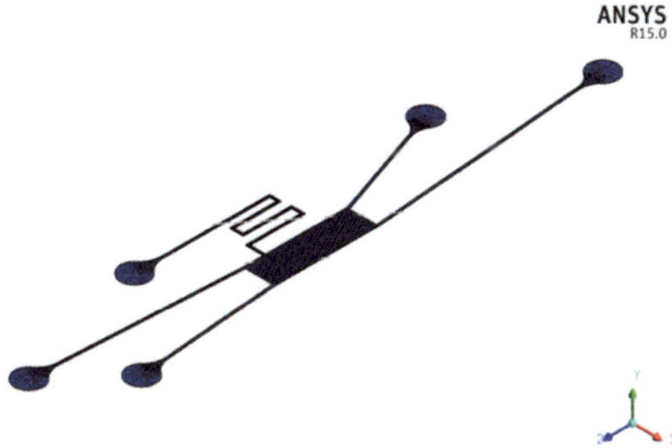

Fig. 7.14 Final design of the microfluidic system

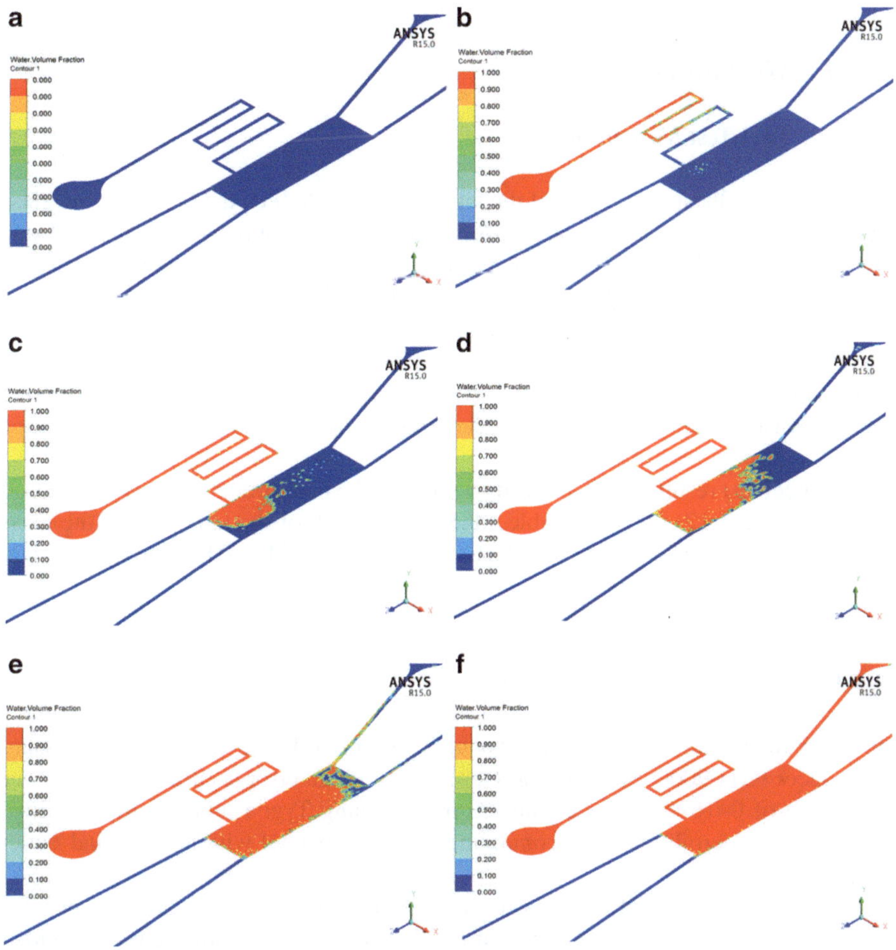

Fig. 7.15 Fluid flow trajectories in the final design at different time frames. (**a**) 0 s, (**b**) 0.1 s, (**c**) 0.2 s, (**d**) 0.3 s, (**e**) 0.4 s, (f) 1 s

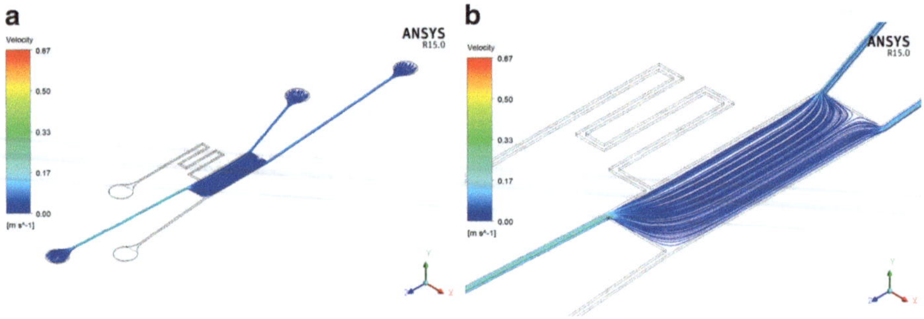

Fig. 7.16 Fluid flow trajectories in the chamber. (**a**) 0 s, (**b**) 0.1 s

7.3.4 Mechanical Stability

7.3.4.1 Peel Test Measurements

To assess the adhesion strength of the tape to either side (i.e., the COC and the glass substrate), a 90°—peel test was performed using a Multipurpose Bondtester (Dage 4000) (Nordson, UK). The test was performed both for a 2.5 cm by 6 cm stripe of tape being bonded to the glass substrate and to the COC substrate. The peeling speed was set at 500 μm/s and the load range at 40 N. After mounting the sample, the sample movement is automatically adjusted to ensure perpendicular peeling force to the surface. The peel strength (N/mm) was calculated by dividing the mean value of the measured peel force (N) in a steady region by the width of the stripe of tape (mm).

The peel test results are visualized by plotting the peel strength versus the elapsed test time (Fig. 7.17). An acceptable adhesion strength of the tape to both the glass substrate (0.18 ± 0.02 N/mm) and the COC substrate (0.21 ± 0.02 N/mm) is shown. (Table 7.1)

7.3.4.2 Burst Test Measurements

In order to measure the pressure at which the microfluidic chip fails, a pressure transducer (CTE9000, First Sensor, Germany) with an error accuracy of 0.2% over the range of 0–16 MPa was connected in-line using Tygon tubing with a syringe pump and a microfluidic chip containing a single channel with an inlet and outlet (Fig. 7.18).

The voltage output of the pressure transducer was monitored using an oscilloscope, and was converted to a pressure value using a calibration curve provided by the manufacturer. Colored dye was pumped through the microfluidic device at different flow-rates, starting at 500 μL/min, and the internal pressure build-up was recorded after the voltage demonstrated by the oscilloscope became stable for at

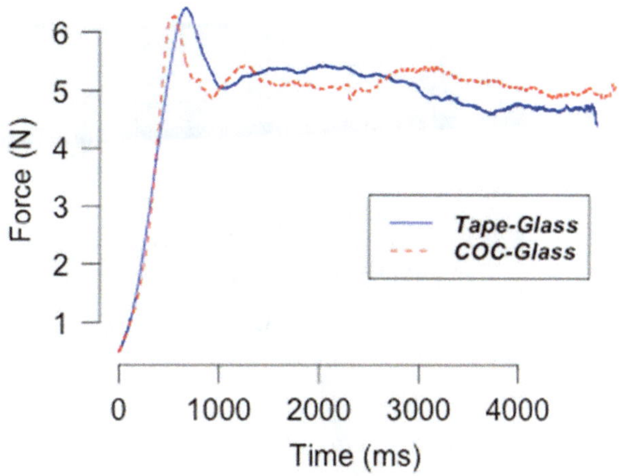

Fig. 7.17 Mean peel strength versus elapsed time of the tape on the glass and the COC substrates (tapewidth = 25 mm)

Table 7.1 Mean peel force calculated for each stripe of tape bonded to the glass substrate or the COC substrate (Tape width = 25 mm)

	Tape–Glass		Tape–COC	
Test #	Mean peel force (N)	N/mm	Mean peel force (N)	N/mm
1	4.69	0.19	5.69	0.23
2	5.22	0.21	5.19	0.21
3	4.68	0.19	4.81	0.20
4	5.06	0.20	4.92	0.20
5	3.63	0.15	4.44	0.18
6	3.82	0.15	6.12	0.24
7	3.77	0.15	4.74	0.19
8	4.07	0.16	5.60	0.22
9	5.28	0.21	5.32	0.21
10	4.74	0.19	5.47	0.22
11	4.81	0.19	5.38	0.22
Mean ± SD	4.52 ± 0.57	0.18 ± 0.02	5.24 ± 0.46	0.21 ± 0.02

least 1 min. As the outlet was open, the pressure at the outlet was considered 1 atm. Taking this in mind, the pressure drop in the channel was calculated.

The burst pressure is defined as the maximum internal pressure recorded prior to device failure due to delamination at the bonding interface or fluidic connections. The burst pressure for each flow-rate is the average of 3 measurements.

The burst test showed the ability of the microfluidic device to sustain typical pressure levels when pumping liquids through the microfluidic channels (Table 7.2).

The measured pressure at the inlet was in concordance with the pressure calculated theoretically (Fig. 7.19). In a rectangular channel, the equation estimating the pressure drop across the channel is expressed as follows [6]: (Eq. (7.2))

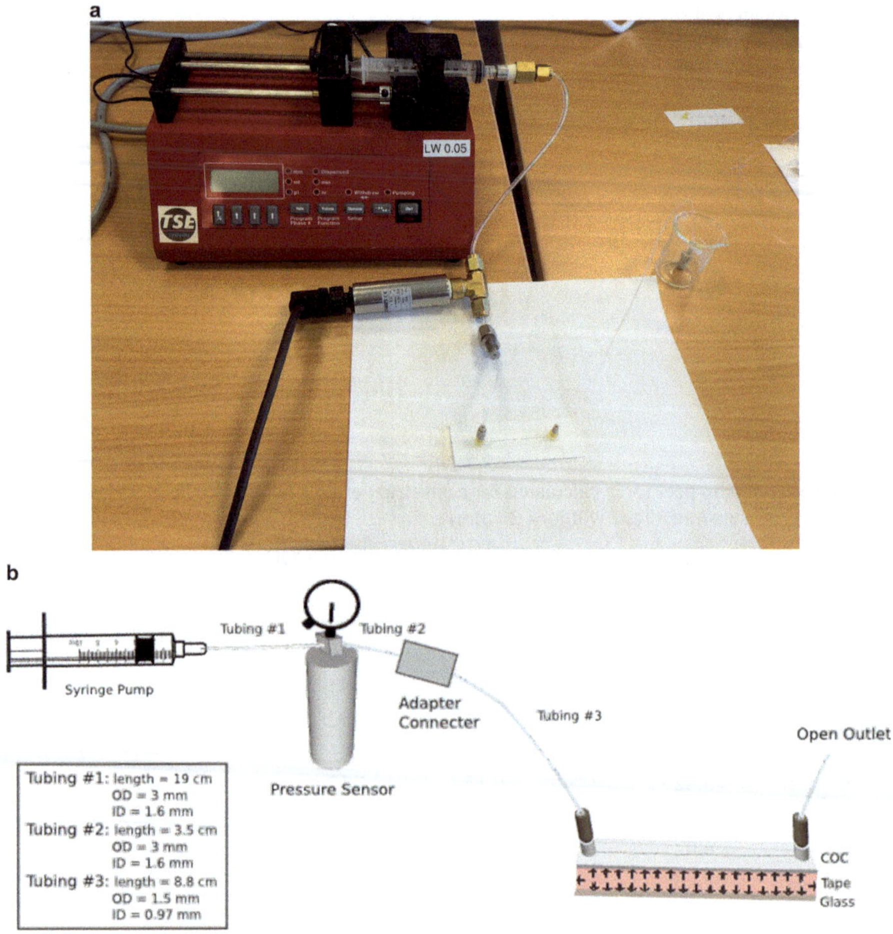

Fig. 7.18 (**a**) Experimental burst test setup, (**b**) Schematic image of the burst test setup

$$\Delta P = Q.R = \frac{12 * \mu * Q * l}{h^3 * w(1 - 0.63 * \frac{h}{w})} \tag{7.1}$$

in which Q is the volumetric flow-rate of the liquid, R is the fluidic resistance, l is the length of the microfluidic channel (39.3 mm), μ is the viscosity of the fluid (10^{-3} Pa s), w is the width of the microfluidic channel (0.264 mm), and h is the height of the microfluidic channel (0.0864 mm).

The experimental burst test results showed that inlet flow-rates higher than 5000 μL/min may fail either at the point of attachment of the interconnect, or at the interface between the two bonded parts.

Table 7.2 Pressure drop calculated for different flow-rates

Flow-rate (μL/min)	Pressure at center (bar)	Pressure drop (bar)
0	–	–
200	1.12	0.12
500	1.30	0.29
1000	1.58	0.58
1200	1.70	0.70
1500	1.87	0.87
1700	1.99	0.99
2000	2.17	1.17
2200	2.28	1.28
2500	2.46	1.46
3000	2.75	1.75
4000	3.33	2.33
5000	–	–

Calculated using Eq. (7.2)

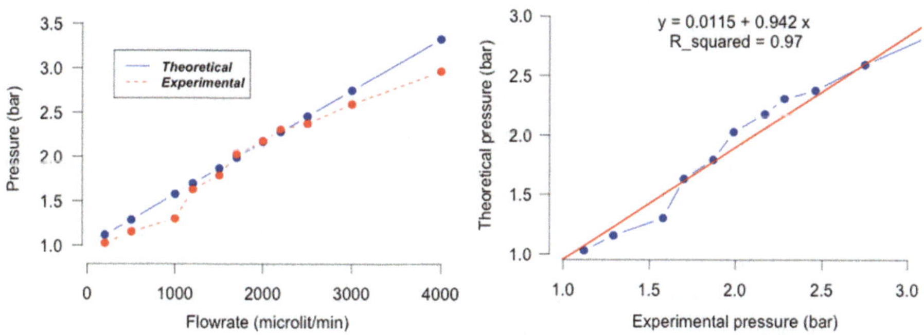

Fig. 7.19 (**a**) Measured vs expected pressure at the inlet for different flow-rates. (**b**) Measured vs expected pressure at the inlet

Pressure drop in the system was calculated as the following: (Eq. (7.1))

$$\Delta P = \frac{8 * \mu * Q * l}{\pi * r^4} \tag{7.2}$$

in which Q is the volumetric flow-rate of the liquid (m^3/s), l is the length of the tubing (m), μ is the viscosity of the fluid (10^{-3} Pa s), r is the radius of the tubing (m).

In order to calculate the highest pressure drop in the system, maximum flow-rate (4000 μL/min $= 0.06\times10^{-6}$ m^3/s) tolerated by the channel was used. Using the parameters shown in Fig. 7.18, pressure drop was estimated to be 0.93 mbar in tubing 1 and 2 and 6.13 mbar in tubing 3. It could be concluded that the pressure drop in the tubing can be neglected compared to the pressure drop inside the microfluidic channel.

7.4 Conclusion

Currently, developing the microfluidic channels and bonding different layers to each other are the bottlenecks in the rapid prototyping process of microfluidic devices. They considerably prolong the turnaround time between the design of the microchannel layout and the fabrication of the final design. We used a cheap, reliable, and rapid method for the fabrication of microfluidic channels using double-sided tapes, enabling not only highly uniform cross-sectional dimensions along the channels but also proper adhesion in hybrid systems, composed of different layers.

Our results showed that the laser-cutting procedure for each sample was performed in 5 min and an additional 5 min was needed for the assembly process. These numbers suggest that the technique is fast and cost-effective compared with the existing microfluidic channel fabrication and bonding techniques. Moreover, modification is not an outlandish task in these devices any more.

The adhesive is shown to have bond strengths that prevent leakage, delamination, and channel rupture in the typical pressure range used for microfluidic applications. A further advantage is that the assembly of the different layers does not require any plasma, heat, or additional treatments and depending on the required precision of the design, can be performed manually.

This intermediate adhesive layer is therefore believed to solve possible obstacles met during the manufacturing of hybrid devices. In other words, while the microfluidic chip described herein has only two levels, a modified version of the fabrication method can be used in applications requiring a larger number of interconnected tapes across additional layers. It should be added that this fabrication method is scalable to larger platforms as it provides sufficient mechanical strength, is reproducible in form and operation, applicable to a broad range of materials, and capable of being produced with high yields.

References

1. J. Hua, S. Wang, L. Wanga, F. Lib, B. Pingguan-Murphye, T. Lub, F. Xua, Advances in paper-based point-of-care diagnostics. Biosens. Bioelectron. **54**, 585–597 (2014)
2. W.B. Hawkins KR, Microfluidic diagnostics for low-resource settings. Proc. SPIE **7593**, 75930L1–75930L15 (2010)
3. A. Nilghaz, D. Wicaksono, D. Gustiono, F. Abdul Majid, E. Supriyanto, M. Abdul Kadir, Flexible microfluidic cloth-based analytical devices using a low-cost wax patterning technique. Lab Chip **12**, 209–218 (2012)
4. D. Duffy, J. McDonald, O. Schueller, G. Whitesides, Rapid prototyping of microfluidic systems in poly(dimethylsiloxane). Anal. Chem. **70**, 4974–4984 (1998)
5. P. Gu, K. Liu, H. Chen, T. Nishida, Z. Fan, Chemical-assisted bonding of thermoplastics/elastomer for fabricating microfluidic valves. Anal. Chem. **83**, 446–452 (2011)
6. S. Mitra, S. Chakraborty, *Microfluidics and Nanofluidics Handbook* (CRC Press/Taylor and Francis, Novato, 2012)

7. B. Flachsbart, K. Wong, J. Iannacone, E. Abante, R. Vlach, P. Rauchfuss, P. Bohn, J. Sweedler, M. Shannon, Design and fabrication of a multilayered polymer microfluidic chip with nanofluidic interconnects via adhesive contact printing. Lab Chip **6**(5), 667–674 (2006)
8. R. Lin, M. Burns, Surface-modified polyolefin microfluidic devices for liquid handling. J. Micromech. Microeng. **15**, 2156–2162 (2005)
9. W. Zhang, S. Lin, C. Wang, J. Hu, C. Li, Z. Zhuang, Y. Zhou, R. Mathies, C. Yang, Pmma/pdms valves and pumps for disposable microfluidics. Lab Chip **9**, 3088–3094 (2009)
10. V. Sunkara, D. Park, H. Hwang, R. Chantiwas, S. Soper, Y. Cho, Simple room temperature bonding of thermoplastics and poly(dimethylsiloxane). Lab Chip **11**, 962–965 (2011)
11. G. Mehta, J. Lee, W. Cha, Y. Tung, J. Linderman, S. Takayama, Hard top soft bottom microfluidic devices for cell culture and chemical analysis. Anal. Chem. **81**, 3714–3722 (2009)
12. N. Steidle, T. Hahn, C. Bader, M. Schneider, R. Ahrens, M. Worgull, A. Guber, Micro and nanostructured microfluidic chip for specific protein immobilization, in 17th International Conference on Miniaturized Systems for Chemistry and Life Sciences, vol. 128 (2013)
13. J. Shah, J. Geist, L. Locascio, M. Gaitan, M. Rao, W. Vreeland, Capillarity induced solvent-actuated bonding of polymeric microfluidic devices. Anal. Chem. **78**, 3348–3353 (2006)
14. J. Steigert, S. Haeberle, T. Brenner, C. Müller, P. Steinert, C.P. Koltay, N. Gottschlich, H. Reinecke, J. Ruhe, R. Zengerle, Rapid prototyping of microfluidic chips in COC. J. Micromech. Microeng. **17**(2), 333–341 (2007).
15. S. Lee, S. Lee, Shrinkage ratio of pdms and its alignment method for the wafer level process. Microsyst. Technol. **14**(2), 205–208 (2008)
16. P. Khashayar, G. Amoabediny, B. Larijnai, M. Hosseini, S. Van Put, R. Verplancke, J. Vanfleteren, Rapid prototyping of microfluidic chips using laser-cut double-sided tape for electrochemical biosensors. J. Braz. Soc. Mech. Sci. Eng. (2016)
17. W. Choi, O. Park, A soft-imprint technique for submicron-scale patterns using a pdms mold. Microelectron. Eng. **73–74**, 178–83 (2004)
18. D. Figeys, Y. Ning, R. Aebersold, A microfabricated device for rapid protein identification by microelectrospray ion trap mass spectrometry. Anal. Chem. **69**(16), 3153–3160 (1997)
19. Z. Zou, J. Kai, M. Rust, J. Han, C. Ahn, Functionalized nano interdigitated electrodes arrays on polymer with integrated microfluidics for direct bio-affinity sensing using impedimetric measurement. Sensors Actuators A Phys. **136**(2), 518–526 (2007)
20. J. Mok, M. Mindrinos, R. Davis, M. Javanmard, Digital microfluidic assay for protein detection. Proc. Natl. Acad. Sci. U. S. A. **111**(6), 2110–2115 (2013)
21. A. Ciftlik, M. Gijs, A low-temperature parylene-to-silicon dioxide bonding technique for high-pressure microfluidics. J. Micromech. Microeng. **21**(3), 035011 (2011)
22. Y. Du, C. Chen, M. Zhou, S. Dong, E. Wang, Microfluidic electrochemical aptameric assay integrated on-chip: a potentially convenient sensing platform for the amplified and multiplex analysis of small molecules. Anal. Chem. **83**(5), 1523–1529 (2011)
23. S. Basrour, L. Robert, X-ray characterization of residual stresses in electroplated nickel used in LIGA technique. Mater. Sci. Eng. A: Struct. **288**, 270–274 (2000)
24. K. Hawkins, M. Steedman, R. Baldwin, E. Fu, S. Ghosal, P. Yager, A method for characterizing adsorption of flowing solutes to microfluidic device surfaces. Lab Chip **7**(2), 281–285 (2007)
25. K. Islam, Y. Jang, R. Chand, S. Jha, H. Lee, Y. Kim, Microfluidic biosensor for beta-amyloid (1–42) detection using cyclic voltammetry. J. Nanosci. Nanotechnol. **11**(7), 6757–6762 (2011)
26. D. Grieshaber, R. MacKenzie, J. Voros, E. Reimhult, Electrochemical biosensors - sensor principles and architectures. Sensors **8**, 1400–1458 (2008)
27. M. Fuerstman, A. Lai, M. Thurlow, S. Shevkoplyas, H. Stone, G. Whitesides, The pressure drop along rectangular microchannels containing bubbles. Lab Chip **7**, 1479–1489 (2007)
28. P. Khashayar, G. Amoabediny, B. Larijani, M. Hosseini, R. Verplancke, D. Schaubroeck, S. Van Put, F. Razi, M. De keersmaecker, A. Adriaens, S. Goemaere, T. Fiers, J. Vanfleteren, A multiplexed microfluidic platform for bone marker measurement: a proof-of-concept. Micromachines **8**(5), 133 (2017)

Chapter 8
Osteokit Validation

Abstract Unstable immobilization of antibody on the electrode surface and non-specific binding are the main factors that determine the detection limit for an immunoassay [3]. It is known that adsorption of proteins onto bulk metal surfaces leads to their denaturation and loss of bioactivity [4]. Therefore, as mentioned earlier, the gold electrodes were modified with AuNPs to overcome these concerns. The AuNP layer also improved the signals because of their large surface area and efficient electron conducting features. This chapter deals with surface modification of electrodes used in Osteokit and its validation process.

We demonstrated the implementation, feasibility, and specificity of this platform (Osteokit) in assaying serum levels of Oc and CTX. In other words, the developed system was sensitive and specific for serum Oc and could detect serum levels of the marker within the range of 2.5–100 ng/mL. This is while the normal reference of the marker is 9–42 ng/mL, suggesting that the system can acceptably detect Oc. Similarly, CTX levels were successfully measured from 1 to 2500 pg/mL. This is while the normal reference of the marker is 50–450 pg/mL, suggesting that the system can acceptably detect CTX.

Our system showed no cross-reactivity for other biomarkers (b-CrossLaps and parathyroid hormone (PTH) for Oc system and Oc and PTH for CTX system). The good correlation between the ECLIA and Osteokit showed that they can be used in the clinically relevant range and other macromolecules available in serum do not affect our results.

8.1 Immunosensing Process

8.1.1 Pretreatment of Working Electrodes

As mentioned in Chap. 5, in order to coat the working electrode with a gold nanoparticle layer that would allow sufficient immobilization of antibodies (Abs) on the surface in later steps, the ECV technique was used.

P. Khashayar, *Point-of-Care Solution for Osteoporosis Management*,
DOI 10.1007/978-3-319-55053-4_8

8.1.2 Surface Modification of Sensing Interfaces

The sensing chip was then bonded with the microfluidic chip, making the device (Osteokit) ready to use. In subsequent steps, the gold surface was modified for capturing antibodies that, during the assay, bind specifically to the analyte molecules (antigens) (Fig. 8.1).

In this regard, the biomarkers were immobilized on the treated surface through covalent binding using cross-linkers [1, 2]. As mentioned in Chap. 6, the electrodes

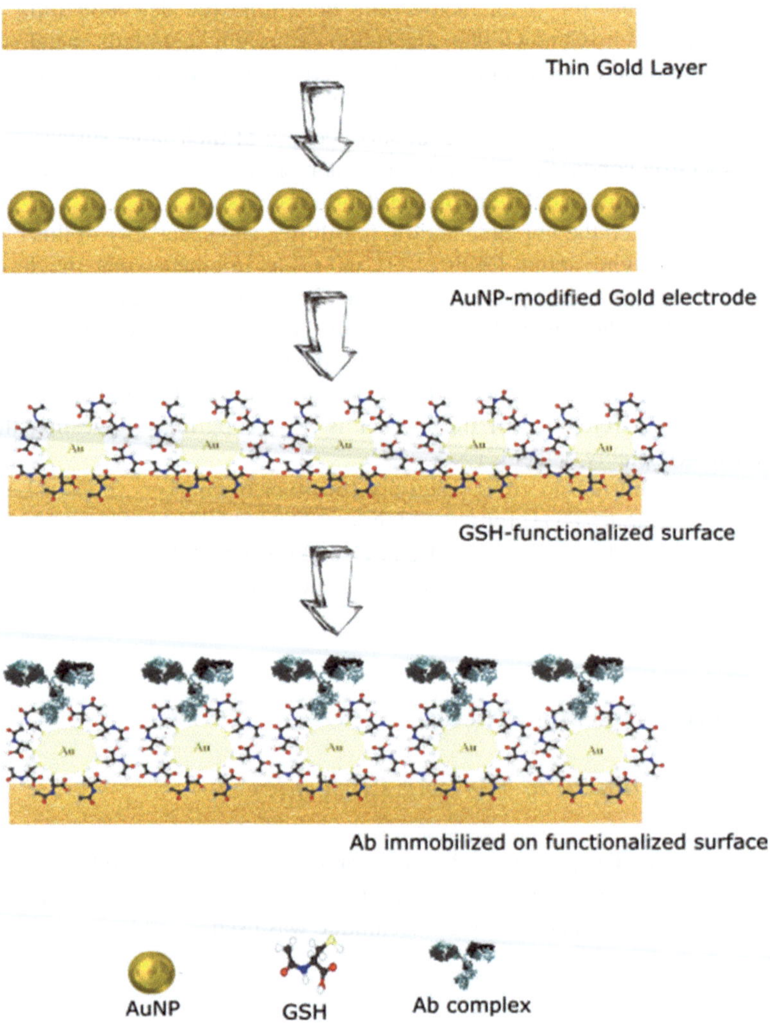

Fig. 8.1 A schematic representation of the immunoelectrode fabrication procedure

were first activated by injecting an aqueous solution of 10 mM GSH via the loading channel to the reaction chamber and held for incubation for 1 h. As the next step, the antibody complex was introduced into the reaction chamber through the inlet microchannel until the chamber was filled with the sample. The sample was then allowed to incubate for 2 h on the sensor electrodes (static condition). The antibody complex was prepared by incubating the antibodies with cross-linkers (100 mM s-NHS for Oc, 400 mM EDC/100 mM s-NHS for CTX) for 2.5 h in room temperature. The cross-linkers provided covalent attractions between the constituents, facilitating the bonding of the primary amine groups (-NH2) of the antibody to the carboxylic group of GSH to form stable amide bonds.

The active parts of the device were washed between each step with DI water to remove unbound fractions. 0.2% (w/v) BSA solution in PBS buffer was then pumped into the microchannel to immerse the treated electrodes for an hour and block unreacted active functional groups. The assembly was kept at room temperature during all steps and finally rinsed with PBS and then with PBS-Tween 20 thoroughly to remove unbound analytes and other non-specific entities from the electrode surface.

8.1.3 Immunosensing Strategy

To investigate the applicability of Osteokit for osteoporosis care, we performed electrochemical immunoassay using fabricated device to measure serum levels of certain BTMs. All experiments were performed within a day to obtain a good sensitivity because the immobilized antibody (Ab) would degenerate a few days after the electrochemical chip is fabricated.

Some of the results of this chapter is published in the article from the author and her team in "Micromachines" [5].

For this purpose, different concentrations of Oc and CTX were applied on the electrodes as the analytes for the sensing test, and the chronoamperometric signals were recorded. The antibody–antigen reaction provides a redox current, which was interpreted by the chronoamperometry. A high quality signal was obtained at +0.65 V in DPV measurements; thus, this potential was taken as the best value for BTM determination.

In this regard, the microfluidic channels were washed using PBS. Thereafter, the detection solution, $K_3[Fe(CN)_6]$, was introduced into the reaction chamber to fill the chamber for chronoamperometry measurements (0.1 mM $K_3[Fe(CN)_6]$ as a redox probe, potential of 0.65 V for 15 s). The measured current at 10 s (when the status in the microchannel was stabilized) was considered as baseline. Then again, the channels were washed using PBS.

At this step, sample solutions with different concentrations of each antigen were loaded into the corresponding chamber and left to incubate in the reaction chamber. The rate of analyte capture by surface-bound antibodies can be very slow because flow in microfluidic devices is nearly always laminar and mixing occurs only by diffusion. The optimum detection time was, thus, calculated to be 5 min. Then, the channel was flushed by passing PBS solutions through the channel twice to ensure complete washing and removal of any unbonded molecules.

The samples were injected sequentially onto the sensing interface and the signal was recorded. After each measurement, a relatively large amount of buffer solution was injected to wash the analyte sample and fill the reaction channel. The specific bonding between the antibody and antigen helps improve both the sensitivity and selectivity of the assay.

The electrodes-on-chip were utilized in these steps. Previous studies have reported a slightly different offset between the working and reference electrodes in sensors with individualized quasi-reference electrodes, similar to these ones, that will consequently affect the potentials. The measured currents were, therefore, calibrated based on the baseline current for each electrode to obtain formal potential of the 3-electrode system. In other words, the current ratio was calculated by dividing the current measured after incubating the electrode with the specific concentration of the antigen by the current measured from the same electrode after the Ab-functionalization process.

As a result, the average of the ratios calculated from several electrodes incubated with certain antigen concentrations was used to plot the calibration curves.[1]

8.2 Osteokit Characterization

8.2.1 Surface Morphology

Since creating AuNP layer is the first step in surface modification, it is critically important to have a well-dispersed nanoparticle layer with uniform size distribution on the surface. The deposition time and scan rate were, therefore, properly controlled to assure sufficient deposition and uniform distribution of AuNPs. According to our results, 60 electrodeposition cycles were needed to produce enough uniformly distributed nanoparticles with an average size of about 20 nm to cover the gold electrodes. We used different surface characterization techniques, including SEM, AFM, and STEM to characterize the morphology of the as-prepared functionalized electrodes. As it can be seen in Fig. 8.2, numerous uniformly distributed nanoparticles were seen on the AuNP-modified gold electrode. After the immobilization of the antibody, these nanoparticles aggregated and became larger in size.

[1]Detailed explanation on how the calibration curve was plotted and the data were extracted from the plot is discussed in Chap. 6.

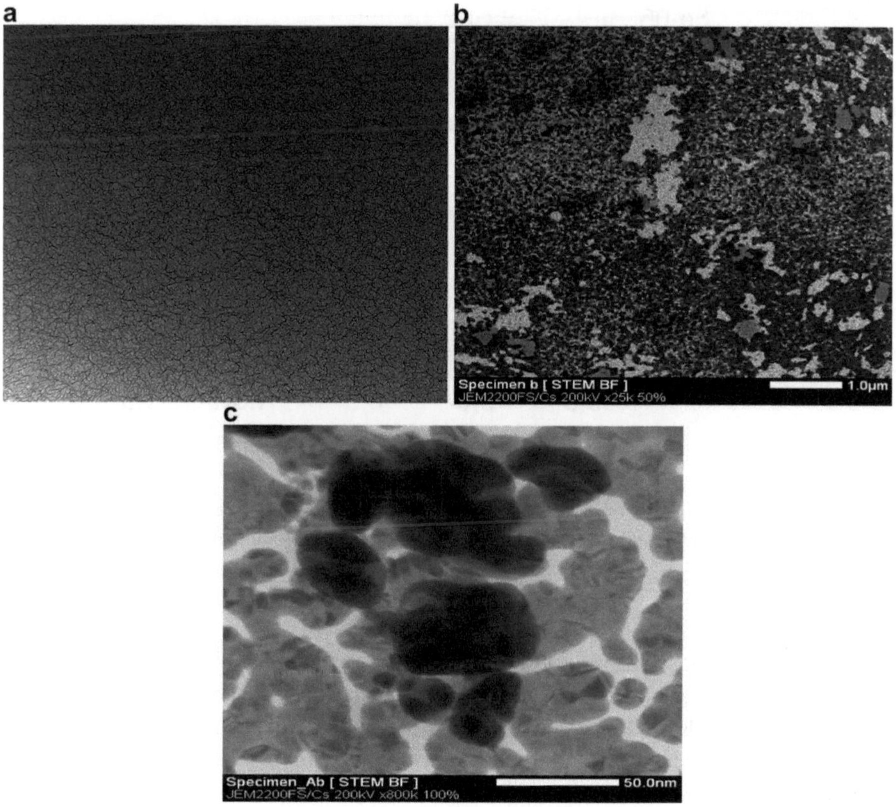

Fig. 8.2 STEM micrographs of (**a**) bare gold nanoparticles by HAADF contrast, (**b**) AuNP-modified gold electrode by BF contrast, (**c**) Ab-functionalized gold electrode by BF contrast (For **a**, scale = 1.0 μm; for **b**, scale = 1.0 μm; for **c**, scale = 50 nm)

8.2.2 Electrochemical Behavior

The electrochemical behavior of the stepwise fabrication of the electrochemical sensor chip was studied using CV, conducted at a potential range from 0 to +1.2 V at 0.1 V/s. The CV curves on the bare Au, after electrodeposition, and after surface modification with Ab in $K_3[Fe(CN)_6]$ are shown in Fig. 8.3. The Au electrode showed well-defined oxidation and reduction behavior of redox moieties Fe(II)/Fe(III) in the electrolyte. The magnitude of response current of the Au electrode was found to increase after the electrodeposition of AuNPs. This suggests the formation of AuNPs and consequent increased electron transfer rate from the medium to electrode.

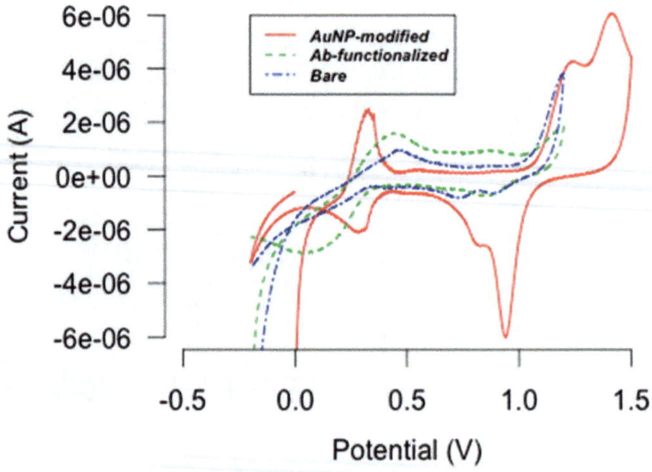

Fig. 8.3 CV measurements of bare (*green*), AuNP-modified (*red*) (External reference and auxiliary electrode was only used for assessing AuNP-modified electrode.), Ab-functionalized (*blue*) electrodes

When antibodies are immobilized on the working electrode, the conductive area of the working electrode is reduced, the electrical resistance is increased and thus the magnitude of the electrochemical response current is further decreased. This suggests successful immobilization of the antibody on the electrode. Furthermore, compared with bare electrodes, the as prepared AuNPs-based electrode-on-chip showed a 7.5-fold larger current response. This obvious signal enhancement should be attributed to the high loading of AuNPs on the gold electrode.

Later on, the presence of antigens over working electrode resulted in an increase in the double layer capacitance and hence reduced electrochemical behavior of test solution proportional to the concentration of the antigen. This was used as the basis for determination of antigen concentration.

Incubation time is an important parameter affecting the immunoassay efficacy. DPV studies revealed that the magnitude of current response decreases over time due to the increased binding between antibody and antigen, causing electron transfer resistance. However, no further decrease in current was observed after 5 min, indicating saturated binding of the immunoreaction. Hence, during the electrochemical sensing measurement the electrode-on-chip was incubated with each antigen concentration for 5 min.

8.3 Calibration Curve

The electrochemical response of Osteokit was studied as a function of different concentrations of Oc and CTX using chronoamperometry technique under identical experimental conditions (0.1 mM $K_3[Fe(CN)_6]$ as a redox probe, potential of 0.65 V for 15 s), respectively. (Fig. 8.4) It is noted that the sample consumption of this prototype chip device is only 100 μL, which is much less than the traditional methods (10 mL).

The magnitude of response current ratio decreased with increasing antigen concentrations; this is due to the formation of immune-complexes between antibody and antigen resulting in electron charge transfer hindrance at the electrode electrolyte interface (Fig. 8.5). The magnitude of current ratio in the Osteokit with CTX

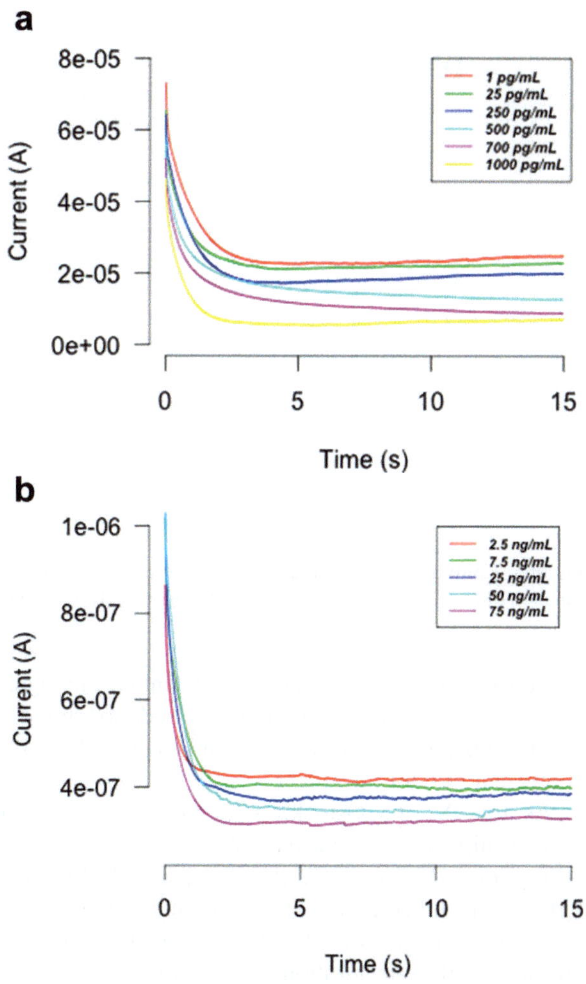

Fig. 8.4 Graphs obtained for the chronoamperometry detection of several concentrations of (**a**) CTX (**b**) Oc using the Osteokit

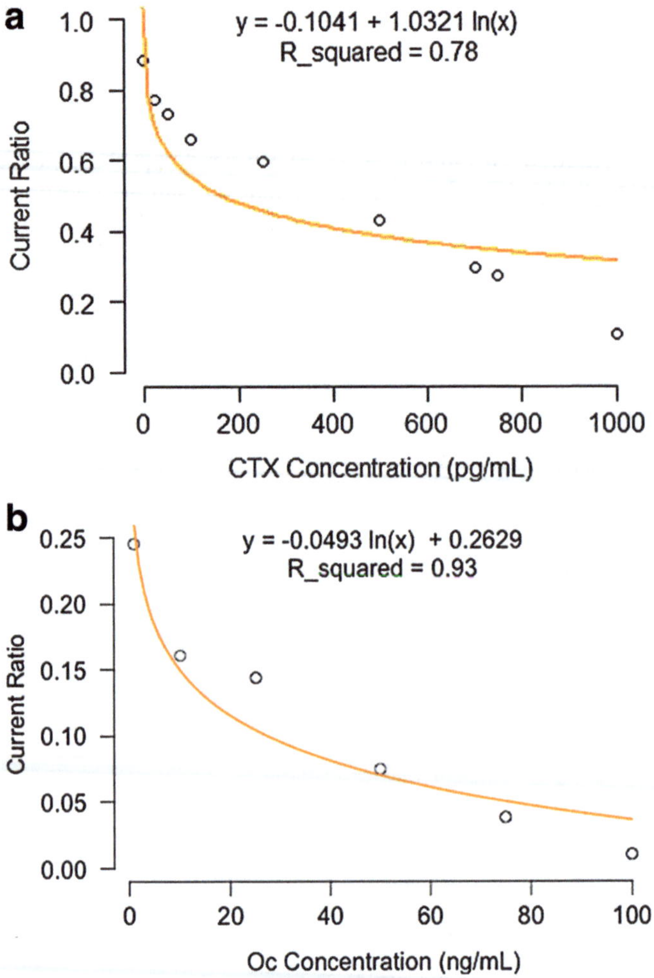

Fig. 8.5 Calibration curve obtained from the electrochemical response studies of Osteokit as a function of (**a**) CTX concentration (1–1000 pg/mL) (**b**) Oc concentration (1–100 ng/mL).

immunoelectrode was linearly dependent on the logarithmic concentration of CTX with an R^2 of 0.78. The immunoelectrode exhibited a near linear range from 1 to 1000 pg/mL, at a LoD of 1.39 pg/mL, and LoQ of 4.22 pg/mL, with a correlation coefficient of 0.99.

As for Oc, the Osteokit with Oc immunoelectrode exhibited a near linear relationship between current ratio and logarithmic Oc concentration from 1 to 100 ng/mL ($R^2 = 0.93$) (Fig. 8.5). The LoD was experimentally found to be 1.94 ng/mL. It had a LoQ of 5.89 ng/mL with a correlation coefficient of 0.96.

Osteokit demonstrated a large dynamic range, which is in the order of the physiological concentration of the studied antigens in the serum. Notably, our titration curve revealed an apparent sublinearity at the low end of the dynamic range for both analytes, which could be attributed to the translation of the actual number of antigens captured to the quantity of AuNPs.

In order to improve the results, new immunoelectrodes with larger work area ($2.5\times2.5\,mm^2$) were used. The magnitude of response current increased with a power of 10 in the Osteokits with new electrodes (Fig. 8.6). Using the new device, the magnitude of current ratio in the Osteokit with the CTX immunoelectrode was linearly dependent on the logarithmic concentration of CTX with an R^2 of 0.96.

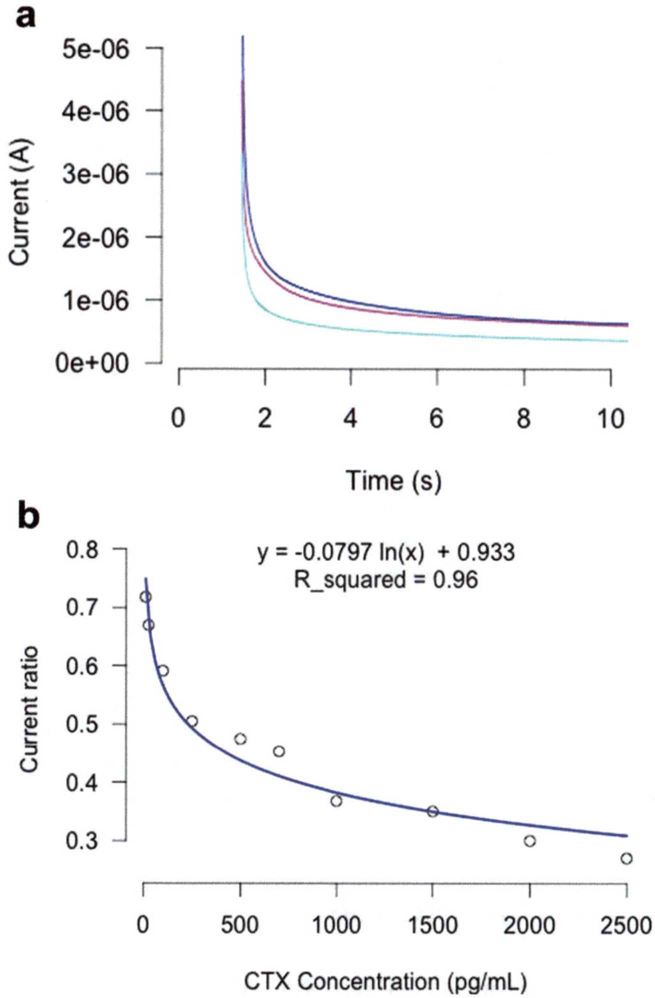

Fig. 8.6 (**a**) Graphs obtained for the chronoamperometry detection of several concentrations of CTX, (**b**) Calibration curve obtained from the electrochemical response studies of Osteokit as a function of CTX concentration (1–2500 pg/mL), using immunoelectrodes with larger work area

The immunoelectrode exhibited a near linear range from 1 to 2500 pg/mL, at a LoD of 2.77 pg/mL, and LoQ of 8.40 pg/mL, with a correlation coefficient of 0.91. It could be concluded that increasing the area of the work area significantly improved the results.

8.4 Selectivity and Cross-Reactivity

Cross-reactivity and selectivity was checked using the same technique reported in Chap. 6. Osteokit showed a minimum change from baseline value (3–5%) with respect to Oc (100 pg/mL) and PTH (100 pg/mL) as interferents on the CTX immunoelectrode. Similarly there was no remarkable response for CTX (100 ng/mL) and PTH (100 ng/mL) on the Oc immunoelectrode. These results show that Osteokit is selective for both CTX and Oc and no cross-talk occurs between the channels.

8.5 Stability and Repeatability

The sensing interface was used ten times without sacrificing its detection efficiency. When the sensing surfaces were regenerated for more tests, the detection of either antigen was not as sensitive as before. This might be a result of losing the activity of the immobilized antibody by repeatedly using the sensing surfaces.

To investigate the stability of the chips, the sensing interfaces were stored at room temperature for a week. Compared with the relative freshly fabricated sensing interfaces, the signal of the conserved sensing interfaces only decreased by 8.1% for Oc and 8.8% for CTX. The small decrease may be ascribed to the partial degradation of the immobilized antibody.

In addition, the results showed a remarkable reproducibility, tested by the application of the same concentration of either antigen on different platforms. They also revealed that the signal obtained in the respective channels is highly reproducible for both CTX and Oc (RSD lower than 8%), indicating an excellent degree of reproducibility for both detection platforms.

8.6 Real Serum Measurements

To compare the sensitivity of our platform with that of current gold standard methods, real samples (here, human serum) previously tested with ECLIA were used as analyte.[2] Again, serum samples of ten patients, whose Oc and CTX levels were

---

[2]Oc samples were obtained from EMRI laboratory and CTX samples were obtained from UZ Hospital.

Table 8.1 Precision of Oc and CTX assay using ECLIA and Osteokit.

Oc samples	ECLIA results (ng/mL)	Osteokit results (ng/mL)
1	27.1	22.7
2	30.8	31.9
3	33.1	29.4
4	25	25.4
5	18.9	18.5
6	15.7	10.6
7	36.9	31.1
8	11.3	18.0
9	17.4	13.8
10	39.7	51.1
CTX samples	ECLIA results (pg/mL)	Osteokit results (pg/mL)
1	1250	1021.6
2	1400	1126.8
3	1070	916.9
4	455	416.9
5	325	284.6
6	2480	1145.9
7	816	812.1
8	92	26.9
9	1670	1132.7
10	194	92.2

recently measured using ECLIA, were used. Using the trend function in Excel and based on the values of two adjacent points on calibration curve, the concentrations of Oc and CTX in the tested samples were calculated according to the obtained current. The results of Osteokit and ECLIA measurements were then compared.

According to the results presented in Table 8.1, the coefficients of variation for the ECLIA and Osteokit were calculated to be 4.6% and 3.7% for Oc, and 6.4% and 7.7% for CTX.

The correlation between our sensor results and that of ECLIA was also investigated and is shown in the correlation and Bland and Altman plots in Figs. 8.7 and 8.8. The R^2 square of the trendline in the correlation plot was 0.85 for Oc and 0.87 for CTX. This indicates that Osteokit has an acceptable correlation with ECLIA in measuring both serum levels of Oc and CTX. These results are more acceptable for lower serum levels of CTX. This could be explained due to the fact that the calibration curve was developed for CTX concentrations between 1 and 1000 pg/mL and the ratio for higher concentrations were extrapolated, which is not really accurate. The sublinearity of the curve at the higher concentrations also aggravates the accuracy issues.

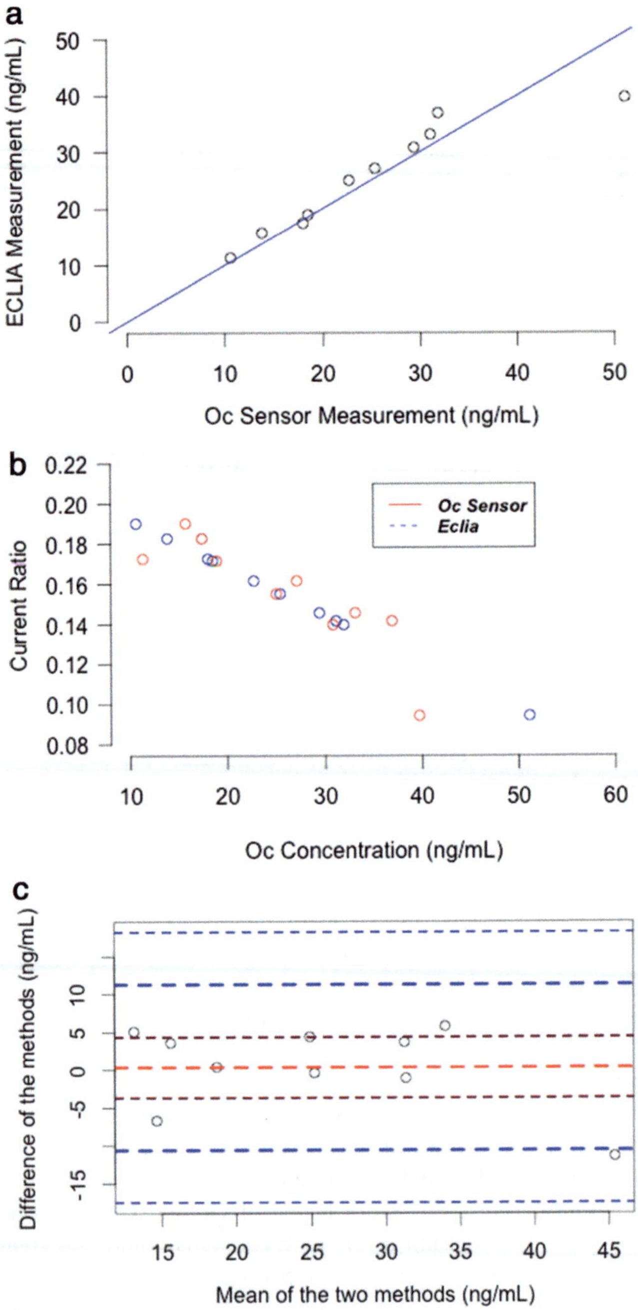

Fig. 8.7 Statistical analysis comparing ECLIA vs. Osteokit in Oc Measurement (**a**, **b**) Correlation plot, (**c**) Bland and Altman plot

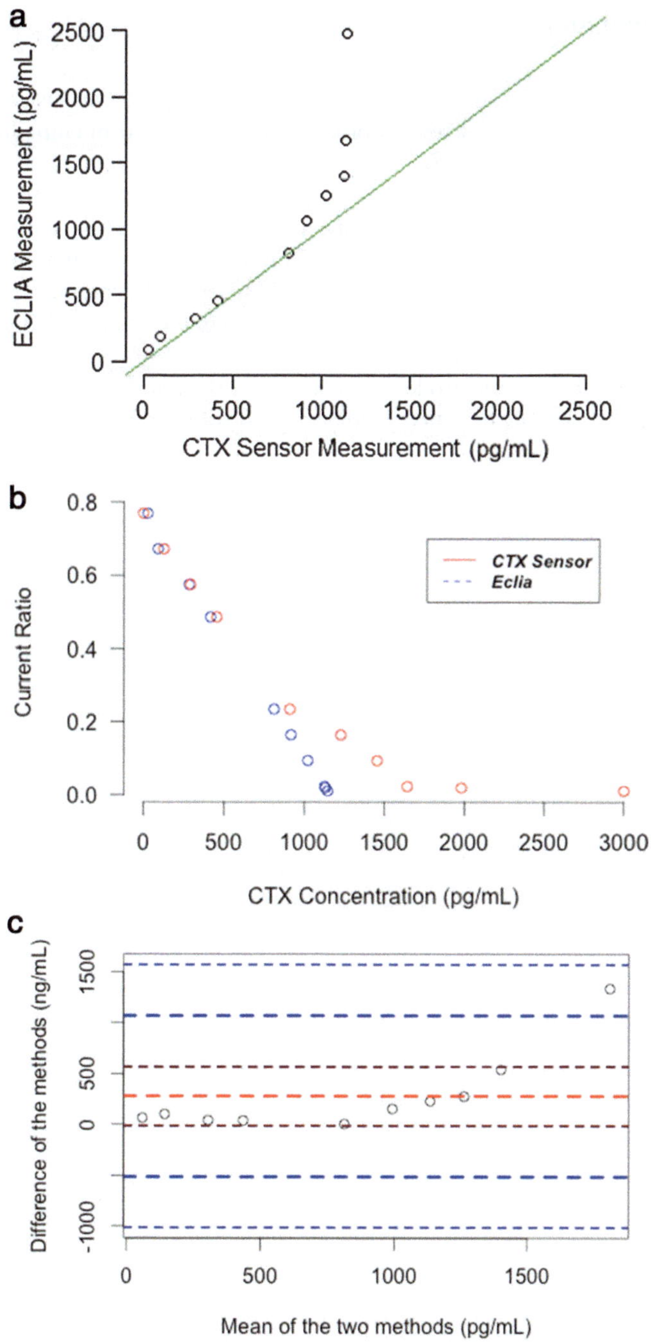

Fig. 8.8 Statistical analysis comparing ECLIA vs. Osteokit in CTX Measurement (**a**, **b**) Correlation plot, (**c**) Bland and Altman plot

8.7 Conclusion

Herein, we presented the development of a microfluidic platform (Osteokit) capable of measuring serum levels of several biomarkers. This platform embodied a unique architecture, which enabled low sample volume requirements, short assay time, simplicity of fabrication and use, and low cost instrumentation and at the same time portability, high sensitivity and specificity for the BTMs. Furthermore, we showed this technology to reproducibly be comparable in sensitivity to the current gold standard ECLIA. As a result, in terms of speed and economic expenses, the feasibility of implementation of such diagnostic tests far from the laboratory could be highly pragmatic.

The application of an electrode with a larger work area in the Osteokit improved all the results, suggesting that this could be a good start for future works in this field.

References

1. J. Křenková, F. Foret, Immobilized microfluidic enzymatic reactors. Electrophoresis **25**, 3550 (2004)
2. B. Brena, P. González-Pombo, F. Batista-Viera, Immobilization of enzymes: a literature survey. *Immobilization of Enzymes and Cells*. Methods in Molecular Biology, vol. 1051 (Springer, New York, 2013)
3. E. Lempens, B. Helms, M. Merkx, Chemoselective protein and peptide immobilization on biosensor surfaces. Methods Mol. Biol. **751**, 401–420 (2011)
4. K. Nakanishi, T. Sakiyama, K. Imamura, On the adsorption of proteins on solid surfaces, a common but very complicated phenomenon. J Biosci. Bioeng. **91**(3), 233–244 (2001)
5. P. Khashayar, G. Amoabediny, B. Larijani, M. Hosseini, R. Verplancke, D. Schaubroeck, S. Van Put, F. Razi, M. De keersmaecker, A. Adriaens, S. Goemaere, T. Fiers, J. Vanfleteren, A multiplexed microfluidic platform for bone marker measurement: a proof-of-concept. Micromachines **8**(5), 133 (2017)

Chapter 9
Final Remark

Abstract To our knowledge, this is the first such device fabricated to measure bone turnover markers (BTMs). Our results also showed the sensitivity of Osteokit to be comparable with the current state of the art, ELISA, and electrochemiluminescence. The developed electrochemical chips showed acceptable sensitivity, specificity, stability, and reliability, thus providing a highly promising potential for clinical applications. The total assay time for this system is about 10 min (loading of antigen, incubation time, flushing with PBS, and testing), while ECLIA needs several hours to be performed.

In this work, only two major BTMs were studied, but the assay system could easily be adapted to other biomarkers in future experiments. These results require further validation but may suggest a direction towards which the field of diagnostic biomarkers is moving.

9.1 Conclusion

The main objective of this book was to develop the first PoC device to be used in osteoporosis care with acceptable sensitivity and specificity, and reduced analysis time. In this perspective, we developed an electrochemical biosensor for BTM measurement in a microfluidic chip architecture for identifying individuals at-risk of fracture or monitoring the treatment process in osteoporotic patients. Our platform consisted of two components: a microfluidic manifold and an electrochemical sensing module.

This fabrication is advantageous in several aspects.

- First, the modification of gold electrode with Au nanoparticles provided a suitable environment for stable immobilization of BTMs keeping their bioactivity in microreactor environment, and thus supporting high quality and low background electrochemical sensing.
- Second, the use of highly specific monoclonal antibodies as bioaffinity sensing interface not only improved the selectivity but also allowed the detection with significantly lower amounts of the sample, compared with conventional methods.
- Third, the use of the microchannels provided controlled functionalization of the sensing platform, while preventing possible contamination and cross-talk.

P. Khashayar, *Point-of-Care Solution for Osteoporosis Management*,
DOI 10.1007/978-3-319-55053-4_9

- Fourth, the combination of the adhesive tape and the milled holder offered a means to rapidly fabricate prototype devices, which could further be tested and the design refined according to the required performance. In other words, the modular design of the chip allows us to easily modify the microfluidic and the electrochemical sensing modules separately.
- Fifth, while serum Oc measurement using traditional techniques needs extra attention,[1] the current PoC device could overcome these concerns.
- Sixth, the final portable setup could be used as a PoC device without the need for trained personnel.
- Seventh, increased assay speed was another advantage of the current device. While a classical sandwich ELISA typically takes 3–5 h, this platform can measure biomarkers in minutes, which is especially important in PoC setting.
- And last, compared with the traditional single-analyte immunoassay, the multiplexed microfluidic immunoassay is more efficient in clinical application since it can quantitatively detect a panel of biomarkers with improved diagnostic specificity.

The most attractive feature of this microfluidic device is the ease and accuracy with which it can deliver information on the clinically relevant parameters that a PoC situation would demand. In other words, this platform is simple in design, inexpensive and easy to fabricate, and at the same time offers accurate measurements. Moreover, the use of microfluidic components along with the employment of the electrochemical detection suggests that the proposed device can be operated as a hand-held device in future.

9.2 Outlook and Future Work

The means to take strategies that can achieve reliable and robust signal amplification, background signal suppression, or non-specific binding prevention is essential for improving the analytical performance of a biosensor. With the rapid emergence of nanotechnology and nanoscience, hybrid bio/nanostructure-based signal amplification holds great promise in realizing high sensitivity and selectivity for the in situ or online detection of biomolecules.

Although this platform has been designed to be adaptable to different types of proteomic assays, initial efforts were focused on the development of a chip to assay two major BTMs, including Oc and CTX. The assay system, however, could easily be adapted to other biomarkers.

---

[1]For accurate measurement of serum Oc, blood samples should be collected at a temperature between 20 and 25 °C, then centrifuged within 90 min; serum may be stored at −20 °C in plastic tubes for up to 26 weeks. Serum may be thawed and refrozen up to 5 times without significant change in measured serum Oc concentration [1, 2].

These results require further validation but may suggest a direction towards which the field of diagnostic biomarkers is moving. Assessing the reproducibility of the results when using real serum and the LSC between Osteokit and ECLIA are needed. Modifying the design in order to measure BTMs in whole blood, using a single inlet for the measurement of all the biomarkers at the same time as well as improving packaging and long-term stability of the electrodes and fabricating a portable read-out are the next steps.

Then, the biosensor must be evaluated in the clinical deployment platform to confirm its performance in predicting or diagnosing the clinical phenotype or outcome of interest as demonstrated in the initial validation phase. It is realistically infeasible to test a large number of patients in this manner due to financial constraints and/or limited availability of patient cohorts. Therefore, the use of prospective-retrospective design and/or biobank samples could be a potential alternative to overcome these obstacles.

References

1. J. Nolla, A. Rozadilla, The time of blood sampling for osteocalcin determinations. Ann. Rheum. Dis. **49**(2), 133 (1990)
2. K. Noonan, M. Kalu, P. Holownia, J. Burrin, Effect of different storage temperatures, sample collection procedures and immunoassay methods on osteocalcin measurement. Eur. J. Clin. Chem. Biochem. **34**(10), 841–844 (1996)

Appendix A
Material and Apparatus

All's well that ends well

Materials and Reagents

Following items were purchased from Sigma-Aldrich:

- Chloroauric acid ($HAuCl_4{\cdot}3H_2O$)
- Sulfo-a-N-hydroxysuccinimide (sulfo-NHS)
- L-Glutathione reduced (GSH)
- Trisodium citrate dihydrate (>99%)
- 1, 8-octanedithiol
- Poly(diallyldimethylammonium chloride) solution (PDDA, Mw 100,000 200,000)
- Potassium hexacyanoferrate (III) ($K_3Fe(CN)_6$)
- Phosphate buffered saline (PBS)
- Bovine Serum Albumins (BSA)
- b-1-ethyl-3-(3-dimethylamonipropyl) carbodiimide (EDC)
- PBS-Tween20
- Albumin-fluorescein isothiocyanate conjugate (FITC-Albumin).

In this work, sulfo-NHS was used instead of NHS as it is more water soluble while showing similar reactivity and specificity.

Carbon (DRP-110) and Gold SPEs (DRP-C223AT) were purchased from Dropsens (Spain). TOPAS 8007 (COC), supplied by Microfluidic Chipshop (Jena, Germany), was used as the upper layer in the Osteokit. Double-coated polyester diagnostic tape 9965 (3M Medical Specialties, USA) was used. It consists of a 0.05 mm opaque white polyester film coated on both sides with a 0.02 mm neutral pressure sensitive acrylate adhesive and supplied between two clear, 0.05 mm silicone-coated polyester release liners. According to the manufacturer, the tape is bioassay, die cut compatible and printable.

Oc monoclonal antibody (ab13418, ab133612) and full-length Oc protein (ab152231) were purchased from Abcam Co. (UK). Anti-collagen type I antibody (MAB1340) and Human Collagen type I (CC050) were obtained from Merck Chemicals (Belgium). In both cases, monoclonal antibodies were purchased despite

P. Khashayar, *Point-of-Care Solution for Osteoporosis Management*,
DOI 10.1007/978-3-319-55053-4

their higher price, as they are more specifically detecting one target epitope and binding with an antigen within a mixture of related molecules extremely efficiently, and less likely to cross-react with other proteins. Moreover, compared with polyclonal antibodies, considering their very high homogeneity, the results are highly reproducible in constant experimental conditions.

PBS was purchased in powder form and was used to prepare a working solution of 10 mM phosphate buffer (NaCl 0.138 M, KCl 0.0027 M), pH 7.4 at 25 °C. PBS and PBS-Tween20 were prepared by dissolving 1 package in 1000 mL of deionized (DI) water. PBS was used to prepare the antibody (10 μg/mL) and antigen solutions.

EDC and sulfo-NHS were dissolved in water at 0.4 M and 0.1 M, divided into small aliquots. EDC aliquots were stored at −20 °C, whereas sulfo-NHS was stored at +4 °C.

The clinical serum samples were obtained from the laboratory of the Endocrinology and Metabolism Research Institute, affiliated with Tehran University of Medical Sciences, and the laboratory of the UZ hospital, affiliated with the Ghent University.

All other reagents were of analytical reagent grade and used without further purification. Distilled water was used throughout the experiments. All the experiments were carried out at room temperature.

Apparatus and Method

All electrochemical experiments were performed using a computer-controlled Dropsens STAT 400 (Dropsens, Spain) during the nanoprobe and Oc biosensor fabrication, and Autolab PGSTAT101 (Metrohm, The Netherlands) for the remaining parts of this work. The tests were carried out in a solution container, at room temperature (23 °C), using the three-electrode configuration fabricated in Chap. 5. Electroactive substance 0.1 mM $K_3[Fe(CN)_6]$, containing 0.01 M NaCl and 0.5 M H_2SO_4, was used as the electrolyte solution.

Cyclic voltammogram (CV) and differential pulse voltammogram (DPV), the main electrochemical testing methods used in the experiments, were performed to confirm surface modification changes and to quantify the target molecule concentration at the sample surface. The CV cycles were performed in 0.1 mM $K_3[Fe(CN)_6]$ solution, containing 0.01 M NaCl, from potentials 0.0 V to +1.2 V, at 0.1 V/s scan rate. CV was also used to determine the formal potential of the electrochemical sensor for DPV. The DPV measurements were performed in 0.1 mM $K_3[Fe(CN)_6]$, containing 0.01 M NaCl, applying following parameters: 0.025 V modulation amplitude, 0.05 s modulation time, 0.005 V step potential and voltage range from 0.5 to 1.1 V.

For the electrodeposition step, external AgCl as reference and Pt. gauze (Alfa Aesar Gmbh & Co, Germany) as auxiliary electrode were used. As for the other steps, gold electrodes-on-chip were utilized.

Electrode surface topography as well as the morphology of the AuNP deposits was measured using Sirion field emission gun scanning electron microscopy (FEG-SEM) (FEI Nova 600 Nanolab Dual-Beam FIB) and atomic force microscopy (AFM). Most high resolution FEG-SEM pictures with magnifications >2.5kx were achieved in "immersion mode" and with the TLD-SE detector (through the Lens Detector). AFM images were acquired on an XE-70 (Park Systems) in non-contact mode ($15 \times 15\,\mu m^2$) using Nanosensors tm (PPP-NCHR).

Scanning transmission electron microscopy (STEM) (Jeol Zeiss—EM10C JEM-2200FS FEG (S)TEM)- 80 operated under high tension of 200 kV) was used to collect information on the prepared gold nanoparticles and to test for their ability to bind with the proteins. The samples were prepared using a gold (Au) TEM grid purchased from TED PELLA, INC (Sweden). A high angle annular dark field (HAADF) detector was used to distinguish the smallest and dispersed prepared gold nanoparticles from the porous gold substrate. A bright field detector (BF) was used to visualize lager prepared nanoparticles.

X-ray photoelectron spectroscopy (XPS) was performed with an electron spectrometer (SSI-SSX100) to estimate the composition of the surface and confirm successful immobilization in each step. The transmission electron microscopy (TEM) images were obtained with an EM10C 80 KV (Zeiss, Germany) transmission electron microscope. The zeta potential was measured by dynamic light scattering (DLS) (MALVERN Instrument MAL1001767 UK), at a temperature of 25 °C and a pH of 7.4. The UV-Vis measurements were carried out on a UV Visible Spectrophotometer SPECORD 250 (Analytik Jena, Germany). Bruker vertex 70 Fourier transform infrared spectroscopy (FTIR) (Germany) was used for further study of the nanoconjugates. A Varian Cary Eclipse fluorescence spectrophotometer equipped with a Varian Cary-block temperature controller was used to assess the protein adsorption capacity of the material.

In order to quantify the uniformity of the cross-sectional geometry of the microfluidic channels along their length, the width and the height of the microfluidic channels were studied at 5 mm intervals using an optical microscope (Nikon Eclipse LV100) and a WYKO NT3300 non-contact optical profiler, respectively.

In order to check the accuracy of the sensor, the electrochemiluminescence immunoassay (ECLIA, Elecsys 2010 autoanalyzer, Roche Diagnostics GmbH, Germany) was used to measure the serum levels of Oc (intra and interassay coefficients of variation of 1.2–4.0% and 1.7–6.5%, respectively) and CTX (intermediate precision CV of <20%).

MIX
Papier aus verantwortungsvollen Quellen
Paper from responsible sources
FSC® C105338

FSC
www.fsc.org

If you have any concerns about our products,
you can contact us on
ProductSafety@springernature.com

In case Publisher is established outside the EU,
the EU authorized representative is:
Springer Nature Customer Service Center GmbH
Europaplatz 3, 69115 Heidelberg, Germany

Printed by Libri Plureos GmbH
in Hamburg, Germany